CIRIA C552

Contaminated land risk assessment
A guide to good practice

D J Rudland

R M Lancefield

P N Mayell

CIRIA *sharing knowledge ■ building best practice*

Classic House, 174-180 Old Street, London, EC1V 9BP UK
TELEPHONE +44 (0)20 7549 3300   FAX +44 (0)20 7253 0523
EMAIL enquiries@ciria.org
WEBSITE www.ciria.org

Summary

CIRIA Research Project 599 reviews the current state of good practice in contaminated land risk assessment, both in the UK and elsewhere. The outcome has been to produce a guidance document and a training pack.

The book sets the context of the risk assessment process within an overall risk management approach. The overall risk management process involves identifying and making decisions concerning risks and subsequent implementation of these decisions. The report describes the stages involved in identifying risks and assessing their significance but stops short of describing remedial actions that might be taken to manage the risk.

The report, and the accompanying training pack (C553) is intended to take the user through the various stages of the assessment process thereby providing guidance to good practice.

The training pack, in seven modules and a workshop study, is primarily intended to be used in a group learning environment, but may also benefit individuals working on their own.

Contaminated land risk assessment. A guide to good practice

Rudland, D J; Lancefield, R M; Mayell, P N

Construction Industry Research and Information Association

CIRIA C552   © CIRIA 2001   ISBN 0 86017 552 9

Keywords	
Contaminated land, risk assessment, hazard, site investigation, risk communication	

Reader interest	Classification	
Geotechnical and environmental engineers, developers, regulators lawyers.	AVAILABILITY	Unrestricted
	CONTENT	Guidance document
	STATUS	Committee guided
	USER	Landowners/developers and their legal advisers, consultants and contractors

Published by CIRIA, 6 Storey's Gate, Westminster, London SW1P 3AU. All rights reserved. No part of this publication may be reproduced or transmitted in any form or by any means, including photocopying and recording, without the written permission of the copyright-holder, application for which should be addressed to the publisher. Such written permission must also be obtained before any part of this publication is stored in a retrieval system of any nature.

Acknowledgements

This report and the accompanying training pack (C553) are the result of CIRIA Research Project 599, "Contaminated Land Risk Assessment – Good Practice Guidance". The research was carried out by Halcrow Group Ltd under contract to CIRIA, and the report was written principally by David Rudland and Robin Lancefield, with support from Patrick Godfrey, Mike Barker and Lindsay Taylor of Halcrow Group, and Peter Mayell of British Aerospace Environmental Services.

The research was guided by a steering group, which comprised:

Dr N J O'Riordan (chairman)	Ove Arup & Partners
Dr B Baker	The Environment Agency
Dr A Barker	FBE Management Ltd
Mr K Deady	ASDA Stores Property Division
Mr J Finnamore	Laboratory of the Government Chemist
Ms V Fogleman	Barlow Lyde & Gilbert
Mr G Fordyce	National House-Building Council
Mr Q Given	London Borough of Camden
Mr G Gray	CIRIA
Mr I Heasman	Taylor Woodrow
Ms S Turney	Railtrack
Mr H Mallett	EnvirosAspinwall
Mr J Navaratnam	English Partnerships
Dr D Nichol	Wrexham County Borough Council
Professor J Petts	University of Birmingham
Mr S Smith	Welsh Development Agency
Dr C Warman	Entec UK ltd
Mr P Williams	Associated British Ports
Mr CDT Wood	Henry Boot Construction (UK) Ltd
Mr P Wood	University of Reading
Mr S Wood	BG Plc

Corresponding member

Ms J Curran   Scottish Enterprise

CIRIA's research manager for the project was Ms J Kwan

The work was funded by:

Department of the Environment, Transport and the Regions through the Partners in Innovation programme
CIRIA's Core Programme
Welsh Development Agency
Environment Agency
English Partnerships
NHBC

CIRIA and the authors gratefully acknowledge the support of these funding organisations and the technical help and advice provided by the members of the steering group. Contributions do not imply that individual funders necessarily endorse all views expressed in published outputs.

Executive summary

Contaminated land has been one of the last major environmental concerns to be taken seriously in the UK and in much of the rest of Europe. It lags far behind air and water quality in terms of statutory and technical approach to control. Until now, contaminated land has been considered mainly in connection with redevelopment of abandoned and derelict land. However, as new UK environmental considerations and related issues such as urban regeneration have come to the fore, this has encouraged consideration of land contamination at times other than redevelopment.

When land is contaminated it can affect human health, the environment and buildings and structures. Contamination affects the uses to which a site can be put and its value. In its worse state, contaminated land can cause unacceptable risk to human health and the environment. Good practice in the management of contaminated land involves assessment of the risk that the contamination might be posing. The history of systematic and consistent risk assessment procedures in the UK is not extensive and is a reflection of both the focus of the legislative regime and the diverse demands of the risk assessment process. This document examines risk assessment and explains the key elements of risk assessment practices and procedures.

PURPOSE AND SCOPE

It is intended that this guidance will be of assistance primarily to those within the construction industry who carry out contaminated land risk assessments on a practical basis with the intention of assisting all practitioners to align their abilities at a common level, in order to promote industry-wide consistency. It will also be of assistance to those who need to know the processes and procedures by which contamination risk assessments are conducted to determine that good practice has been followed.

To prepare this guidance, the practice of contaminated land risk assessment both in the UK and overseas has been reviewed. It is not intended that the document should be used as a comprehensive manual for carrying out risk assessment. Rather, the reader is directed to sources of further information and guidance, which will assist when developing a risk assessment strategy.

The report:

- aims to ensure consistent approach to risk assessment, reflecting current "good practice" and taking into account the development of UK government policy and the rapidly changing legislative regime
- will enable all the interested parties to appreciate the basic processes and procedures involved in risk assessment
- will be applicable to a range of different objectives and types of contaminated sites particularly those being proposed within the construction industry for redevelopment.

Although the procedures and processes are intended to address all types of contaminated land there are instances where specialised approaches may be adopted. Within the report there are sections covering risk assessment of sites with particular types of contamination, such as risk assessment of sites affected by soil gases and radiation.

STRUCTURE AND CONTENT

The report sets the context of the risk assessment process within an overall risk management approach. The overall risk management process involves identifying and making decisions concerning risks and subsequent implementation of these decisions. This report describes the stages involved in identifying risks and assessing their significance, but stops short of describing remedial actions that might be taken to manage the risk.

The outline structure of this report is shown below:

Chapter 1   *Introduction.* Describes the importance of risk assessment, statutory and practical motivation for conducting assessments and the parties involved.

Chapter 2   *The basic framework for the assessment.* The risk-based approach in the UK, and the conceptual stages.

Chapter 3   *Defining risk assessment objectives.* Setting the context and objectives of the assessment.

Chapter 4   *Gathering information for Phase 1 assessment.* Describes the information that it is necessary to gather for Phase 1 assessment and how to apply this to create a conceptual model.

Chapter 5   *Site investigations to acquire data.* Provides an overview of the site investigation data collection process. This topic is well covered by several recent publications, including those by the British Standards Institution, and the reader is directed to these for detailed information.

Chapter 6   *Phase 2 estimation and evaluation – the significance of risk.* Describes how all the data is assessed to determine the significance of the risk.

Chapter 7   *Risk communication.* The situations and methods by which risk is communicated to interested parties.

The appendices provide additional background information on particular topics and provide references for further reading. Some case studies are also presented.

THE USERS OF THIS REPORT

Decisions involving potentially contaminated land tend to involve a diverse range of interested parties. At one end of the scale there are regulators and planners, each with the objective of protecting the environment. At the other end are pressure groups and members of the public who will have limited experience of the technical issues of contaminated land but nevertheless will have their own perceptions of the problems involved. Landowners, developers and their financial backers aim to complete the development mindful of costs versus benefits. Consultants may provide advice to all groups, who nevertheless will require sufficient understanding to be able to make informed decisions. These parties will have a broad range of interests, understanding and technical ability, and may enter the decision process at different stages.

It is against this background that the guidance is intended to assist all these parties to be able to contribute to the consultation and agreement through to the remedial design and validation stages that are commonly involved in contaminated land assessment and development. It is assumed that users of this report will have had some involvement in contaminated land assessment and will most probably be planning to carry out their own risk assessments. Alternatively the user may be required to review and assess the results of another's assessment.

RELATIONSHIP TO OTHER DOCUMENTS

This report is intended to take the user through the various stages of the assessment process, identifying the key stages and providing pointers towards best practice. The volume is self-contained to some extent, but where further reading is recommended the user will be directed to other documents that will provide them with the specialist background in the subject of concern. Some of these will discuss detailed scientific and technical aspects whilst others will discuss procedural approaches. The report draws particularly upon contemporary guidance that is produced in a regulatory context. In particular the report draws upon the UK Department of the Environment, Transport and the Regions' document *Handbook of Model Procedures for the Management of Contaminated Land* (Contaminated Land Research Report CLR11).

The accompanying Training Pack has been designed to help those involved in the management of contaminated land understand the procedures involved in assessing risks due to contamination in a variety of contexts.

THE NEED FOR SPECIALIST ADVICE

The assessment of contaminated land is a specialist activity that is likely to involve at some stage a significant technical input. It is not uncommon for decisions involving contaminated land to require specialist technical advice from biologists, chemists, engineers, environmental scientists, toxicologists and others. The guidance will not be a substitute for the professional advice that will be required in many cases, particular where specialist technical skills are required.

Contents

Summary 2
Acknowledgements 3
Executive summary 4
List of figures 10
List of tables 11
List of boxes 12
Glossary 13
Abbreviations 19

1 INTRODUCTION 21
    1.1 What is contaminated land? 21
    1.2 Why is contaminated land important? 21
    1.3 Why then consider contaminated sites at all? 21
    1.4 How much contaminated land is there? 22
    1.5 Why is contaminated land risk assessment important? 22
    1.6 The purpose of this report – who should be interested in this guidance? 22

2 THE BASIC FRAMEWORK FOR CONTAMINATED LAND RISK ASSESSMENT 25
    2.1 Introduction 25
    2.2 The risk management approach 25
    2.3 The risk-based approach in the UK 25
    2.4 The main stages of risk assessment 26
    2.5 Further reading 30

3 DEFINING RISK ASSESSMENT OBJECTIVES 31
    3.1 Introduction 31
    3.2 Deciding the context 31
    3.3 The consequences of risk assessment 32
    3.4 Agreement of the scope of the assessment 33
    3.5 Proceeding with the assessment 34

4 PHASE 1 RISK ASSESSMENT – GATHERING INFORMATION 35
    4.1 Why is the information-gathering stage necessary? 35
    4.2 How is the information gathered? 36
    4.3 The steps of Phase 1 assessment 36
        4.3.1 Step 1 – define boundaries 36
        4.3.2 Steps 2 and 3 – obtain and collate information 36
        4.3.3 Step 4 – pollution linkages 40
        4.3.4 Step 5 – developing the conceptual contaminant-pathway-receptor model 42
        4.3.5 Step 6 – drawing conclusions and fulfilling objectives 45
    4.4 Reporting 45
    4.5 Further reading 46

5 PHASE 2 RISK ESTIMATION – SITE INVESTIGATIONS TO ACQUIRE DATA 47

5.1 Purpose of this chapter 47
5.2 How site investigations enhance risk assessment 47
5.3 Confidence in the data 48
5.4 Developing an investigation strategy 49
    5.4.1 Introduction 49
    5.4.2 Staging site investigations and "zoning" 50
    5.4.3 Consultations 50
    5.4.4 Health and safety 51
    5.4.5 Site investigation techniques 51
    5.4.6 Sampling 53
    5.4.7 Laboratory analysis 55
5.5 Conclusions 56
5.6 Further reading 58

6 PHASE 2 – ESTIMATION AND EVALUATION: THE SIGNIFICANCE OF RISK 61

6.1 Introduction 61
6.2 Risk estimation 63
    6.2.1 Humans – commonly encountered complaints 65
    6.2.2 Humans – landfill gases and other bulk gases 72
    6.2.3 Humans – asbestos 72
    6.2.4 Humans – biological hazards 73
    6.2.5 Humans – explosives and munitions 73
    6.2.6 Humans – radioactive materials 73
    6.2.7 Water environment – all contaminants 73
    6.2.8 Flora and fauna – all contaminants 74
    6.2.9 Buildings materials and service – all contaminants 75
    6.2.10 Risk assessment models 76
6.3 Risk evaluation 79
    6.3.1 Collating and reviewing risk-based information 79
    6.3.2 Addressing uncertainty 79
    6.3.3 Identification of unacceptable risks 79
6.4 Reporting 84
6.5 Further reading 84

7 RISK COMMUNICATION 87

7.1 Introduction 87
7.2 Why is risk communication an essential part of the risk assessment process? 87
7.3 What are the different perceptions of risk? 88
7.4 How to communicate contaminated land risks 90
7.5 Communicating uncertainties 92
7.6 Communication of risk to other assessors and remedial contractors 93
7.7 Conclusion 94
7.8 Further reading 95

REFERENCES 97

A1	**PUBLICATIONS ON CONTAMINATED LAND FOR FURTHER READING**	**103**
	A1.1 UK Government publications – general publications	103
	A1.2 Reports sponsored by UK Department of Environment, Transport and the Regions (DETR) and other government departments	104
	A1.3 Environment Agency publications	108
	A1.4 BRE (Building Research Establishment) publications	110
	A1.5 Construction Industry Research and Information Association (CIRIA) publications	110
	A1.6 Health and Safety Executive publications	111
	A1.7 Other publications	112
A2	**SUMMARY OF THE LEGISLATIVE REGIME IN THE UK AND OTHER COUNTRIES**	**113**
	A2.1 The UK regime	113
	A2.2 The European Union	115
	A2.3 Other countries	115
	A2.4 Further reading	115
A3	**CONTAMINANTS REQUIRING SPECIALIST ADVICE**	**117**
	A3.1 Introduction	117
	A3.2 Land contaminated by biological hazards	117
	A3.3 Soil gases	119
	A3.4 Radioactive hazards	122
	A3.5 Munitions and explosives	124
	A3.6 Asbestos	127
A4	**ECOLOGICAL RISK ASSESSMENT**	**131**
	A4.1 What are ecological risk assessments?	131
	A4.2 Why are ecological risk assessments undertaken?	131
	A4.3 The scope of an ecological assessment	132
	A4.4 How to carry out an ecological assessment	132
A5	**ESTIMATING RISKS TO BUILDING FABRIC AND SERVICES**	**137**
	A5.1 Why assessing risks to the building fabric and structures is important	137
	A5.2 What materials are concerned?	137
	A5.3 Further reading	142
A6	**RISK ASSESSMENT SOFTWARE MODELS**	**145**
	A6.1 BP RISC	145
	A6.2 CONSIM	146
	A6.3 RBCA Toolkit for Chemical Releases	147
	A6.4 RISC-HUMAN	149
	A6.5 Risk* Assistant	149
A7	**CASE STUDIES**	**151**
	A7.1 Case study 1 – housing development, Liverpool, UK	151
	A7.2 Case study 2 – contamination of groundwater at maintenance facility	152
	A7.3 Case study 3 – land purchase by holding company, north-west England	154
	A7.4 Case study 4 – large brownfield site redevelopment, Essex, UK	155
	A7.5 Case study 5 – small housing development, UK	157

List of figures

Figure 1.1	Contaminated land stakeholders	25
Figure 2.1	The risk management process	27
Figure 2.2	Summary of the risk assessment process	29
Figure 2.3	Process for risk assessment	30
Figure 3.1	The sequential process for risk assessment	34
Figure 5.1	Soil gas monitoring standpipe apparently indicating ambient air-gas concentrations	57
Figure 5.2	Example of an appropriate soil gas-monitoring standpipe	57
Figure 6.1	Estimation and evaluation of risk from site investigation data	62

List of tables

Table 2.1	Receptors, for the purposes of the Environmental Protection Act 1990	28
Table 3.1	Examples of technical and non-technical considerations	33
Table 4.1	Some key sources of information and their usefulness	39
Table 4.2	Overview of selected contaminants and associated hazards	41
Table 4.3	Example of a preliminary conceptual model in tabular form	44
Table 5.1	A general guide to the necessity for testing at the location of sampling	54
Table 6.1	The exposure pathways considered in CLEA	68
Table 6.2	Summary of the principles and methodologies of selected risk assessment models	77
Table 6.3	Classification of consequence	80
Table 6.4	Classification of probability	80
Table 6.5	Comparison of consequence against probability	82
Table 6.6	Description of the classified risks and likely action required	82
Table 7.1	Perception of the public compared with that of experts	88
Table 7.2	Lay criteria for judgement of science	90

List of boxes

Box 3.1	Examples of risk management objectives	32
Box 3.2	Example of a decision based on risk assessment	33
Box 3.3	Tips for good practice	34
Box 4.1	Information that is normally required for Phase 1 assessments	37
Box 4.2	Example illustrating the determination of pollutant linkages	42
Box 4.3	Risk assessment in relation to development and construction activities	44
Box 4.4	Content of a typical Phase 1 report	46
Box 4.5	Tips for good practice	46
Box 5.1	Consequences arising from incomplete identification of pollutant linkages	48
Box 5.2	Sample size versus total soil mass	49
Box 5.3	Items for inclusion in the site safety plan	51
Box 5.4	Items to be considered when choosing site investigation techniques	52
Box 5.5	Tips for good practice	56
Box 6.1	Examples of risk estimation without site investigation data	63
Box 6.2	Comparison of site investigation data to assessment criteria	64
Box 6.3	Illustration of ecotoxicological and human health consideration within Dutch threshold and intervention values	67
Box 6.4	Risk estimation of mixtures of substances	69
Box 6.5	Terminology	70
Box 6.6	Tolerable daily soil intake	70
Box 6.7	The assessment of carcinogenic compounds	71
Box 6.8	Bioavailabilty	72
Box 6.9	Assessment criteria – water resource protection	74
Box 6.10	Example of risk evaluation	83
Box 7.1	Example of community concerns over a contaminated site	91
Box 7.2	Example of mistiming the risk communication process	91
Box 7.3	Tips for good practice	94

Glossary

acceptable daily intake (ADI)	An estimate of the daily exposure dose that is likely to have no harmful effect even if continued exposure occurs over a lifetime
acceptance test	A statistical test used to decide how a set of soil analytical data compares with a generic (or comparable) site-specific assessment value
aromatic	A hydrocarbon compound containing a benzene ring structure
assessor	An individual, or in some cases a team, instructed to carry out the risk assessment for a site or number of sites.
attenuation	The process by which a compound (or pollutant) is reduced in concentration over time, through absorption, adsorption, degradation, dilution, and/or transformation.
autocorrelation	Extent to which contaminant concentrations in soils appear spatially related, as opposed to randomly distributed
averaging area	An area within which a human receptor may be exposed to hazardous substances; the size of the area depends on the actual or intended use of the site
brownfield sites	Any land that has been previously developed or requires work done to it to bring it into use
cancer potency factor	See *slope factor*
chronic risk	The probability that an adverse effect will occur as a result of long-term exposure to, or contact with, a hazardous substance or as a result of a long-term hazardous condition
clarity	Extent to which the available information presents a clear and unambiguous account of the situation being assessed
completeness	Extent to which the available information adequately describes the characteristics of contaminants, pathways and receptors
conceptual model	A textual or graphical representation of the relationship(s) between contaminant(s), pathway(s) and receptor(s) developed on the basis of Phase 1a risk assessment findings, and refined during subsequent phases of assessment
conservatism	Extent to which risk assessment models and assumptions take a precautionary approach to human health and environmental protection
contaminant	See *source*

contaminated land (for the purposes of Part IIA of the Environmental Protection Act 1990)	Any land that appears to the local authority in whose area it is situated to be in such a condition, by reason of substances in, on or under the land, that: a) significant harm is being caused or there is a significant possibility of such harm being caused, or b) pollution of controlled waters is being, or is likely to be caused
contaminated land in assessment of land contamination effects	Land that represents an actual or potential hazard to health or the environment as a result of a current or previous use
contaminated site	Any site that, as a result of activities either previously or currently carried out on it, contains concentrations of substances or pathogens high enough to be a hazard to health or the environment either in the current use of the site or if it is used for a different purpose (Royal Commission on Environmental Pollution 1996 Sustainable use of Soil 19th report)
critical soil concentration value (C_{Scrit})	The unknown soil concentration value, usually designated as the site-specific assessment criteria. This value is calculated during the site specific assessment criteria generation process
decision criteria	Factors taken into account when arriving at particular decisions or judgements
decision summary sheet	A summary record of the decisions made at key stages of risk assessment and of the level of information on which the decisions have been made
desk study	Interpretation of historical, archival and current information to establish where previous activities of the land were located, and where areas or zones containing distinct and different types of soil contamination can be expected to occur, and to understand the environmental setting of the site in terms of pathways and receptors
detailed investigation	Main stage of on-site investigation involving sampling and analysis to characterise ground conditions for a specified purpose; may be undertaken in a single or a number (eg Stages 1 and 2) of successive stages
effective concentration	Concentration of a substance that causes a defined magnitude of response in a given system. EC50 is the median concentration that causes 50 per cent of maximal response.
estimated daily intake (EDI)	The intake, or dose, of a contaminant from a site for a relevant pathway
excess cancer risk	The additional risk an individual has of developing cancer in addition to other non-specified causes
exploratory investigation	Limited intrusive/analytical work carried out to provide preliminary information on the condition of the land

exposure dose	Amount of a substance (chemical, radiological or physical agent) that is available for absorption and is absorbed into the body
flora and fauna	Plants and animals including livestock (agricultural and game species), crops and plants used for landscape and amenity
generic assessment criteria	Criteria derived and published by an authoritative body which take into account generic assumptions about the characteristics of contaminants, pathways and receptors and which are designed to be protective in a range of defined conditions
genotoxic	See *mutagen*
greenfield site	An area previously undeveloped and therefore undisturbed with a predominantly consistent subsurface
harm	Harm to the health of living organisms or other interference with the ecological systems of which they form part and in the case of man, includes harm to their property (Section 78A of the Environmental Protection Act 1990
hazard	A property (of a substance) or situation with the potential to cause harm
hazard assessment	Consideration of the plausibility of pollutant linkages and determination of the potential for risks to human health and the environment
hazard identification	Identification of contaminant contaminants, pathways and receptors taking into account the actual or intended use of the site and its environmental setting
hazard index	The sum of hazard quotients. Represents the effects of projected intakes of chemicals by comparing to toxicity values (reference doses)
hazard quotient	Ratio of chronic daily intake to a reference dose (a ratio of <1 means that the systemic effects are assumed not to be of concern, >1 means that they are assumed to be of concern)
hotspot	A defined area or volume of ground containing elevated concentrations of hazardous substances
leachate	Liquid that has percolated through solid waste and has extracted, dissolved or suspended materials from it
lethal dose	When noted as, say, LD_{50}, indicates the lethal dose required to kill 50 per cent of exposed organisms
made ground	Material artificially in place comprising a wide range of materials such as, concrete, tarmacadam, brick materials
maximum contaminant level	The maximum permissible level of a contaminant in water delivered to any user of a public system

maximum exposure level	Legally enforceable limits acting as safety factors for general exposure indicating the point beyond which exposure will cause harm to human health
mean daily intake (MDI)	A measure of the background intake (in $\mu g \, d^{-1}$) of a contaminant from ambient concentrations in food, water and air for the UK population.
minimum reporting requirements	Minimum amount of information it is considered reasonable to provide when describing the rationale for, conduct of and findings at each stage of risk assessment
model procedures	DETR handbook of model procedures for the management of contaminated land (CLR11)
modifying factor	Applied to the safety factor (when deriving reference doses) to account for quality of data
mutagen	A carcinogen that can induce a genetic alteration in a single cell which may lead eventually to tumour initiation
NOEC/NOAEL	A dose below which no adverse effect is observed, and/or an estimate of the dose level below which there is no adverse effect
occupational exposure standards	Legally enforceable limits related to occupational exposure via the inhalation pathway only
pathway	The means by which a hazardous substance or agent comes into contact with, or otherwise affects a receptor
Phase 1a risk assessment	A discrete phase of risk assessment that incorporates the conceptual stage of hazard identification
Phase 1b risk assessment	A discrete phase of risk assessment that builds on Phase 1a risk assessment findings and incorporates the conceptual stage of hazard assessment
phase 2 risk assessment	A discrete phase of risk assessment which builds on Phase 1a and Phase 1b risk assessment and incorporates the conceptual stages of risk estimation and risk evaluation
pica child	A child who deliberately and habitually ingests soil, or who frequently mouths soil contaminated toys etc
pollutant linkage	The contaminant pathway receptor relationship
polycyclic aromatic hydrocarbon	A hydrocarbon compound containing a fused benzene ring structure
potency slope	See *slope factor*
potentially contaminated land sites	Sites identified (while undertaking desk studies/site investigations) as having been or are subject to a land use that may give rise to contamination).

probability	The likelihood of an event occurring, expressed as a numerical ratio, frequency or per cent
RAMSAR site	Wetland of international importance, especially as waterfowl habitat. Designated under the Ramsar Convention on Wetlands of Importance 1971 (Ramsar Convention), which places general and special obligations on contracting parties relating to the conservation of wetlands throughout their territory
Risk-based corrective action (RBCA)	Derived by ASTM Standard Provisional Guide for RBCA PS104-98 and Standard Guide for RBCA Applied at Petroleum Release Sites E1739-95. Uses a three-stage procedure for dealing specifically with petroleum-contaminated sites
reasonable minimum information requirements	Minimum amount of information it is considered reasonable to obtain to make the relevant (technical) judgements at each stage of risk assessment
reasonably practicable	Reasonably practicable taking into account technical feasibility, increased risk, cost and/or time and whether these are proportionate to anticipated benefit
Rebecca	Jargon for RBCA (see Risk-Based Corrective Action)
receptor	The entity (eg human, animal, water, vegetation, building services etc) which is vulnerable to the adverse effects of the hazardous substance or agent. May also be called the "target"
reference concentration (inhalation exposure)	An estimate (with uncertainty spanning perhaps an order of magnitude or greater) of an exposure level for the human population, including sensitive sub-populations, that is likely to be without an appreciable risk of deleterious effects during a lifetime or prolonged period. Expressed in mg/m^3 (USEPA)
reference dose (oral or dermal exposure)	An estimate (with uncertainty spanning perhaps an order of magnitude or greater) of an exposure level for the human population, including sensitive sub-populations, that is likely to be without an appreciable risk of deleterious effects during a lifetime or prolonged period. Expressed in mg/kg/day (USEPA)
relevance	Extent to which the available information is relevant to the contaminants, pathways and receptors being assessed
reliability	Extent to which measurements or observations accurately reflect the true or likely site conditions taking into account the implications of any gaps in the information
risk	The probability that due to a hazard an adverse effect due to a hazard will occur under defined conditions
risk assessment	Identification, estimation and evaluation of risks
risk estimation	Estimation of the risk(s) that identified receptor(s) will suffer adverse effects if they come into contact with, or are otherwise affected by, contaminant sources under defined conditions

risk evaluation	Evaluation of the need for risk management action having regard to the nature and scale of risk estimates, any uncertainties associated with the assessment process and cost/benefit
risk management	The decision-making process to decide the most appropriate form of remedial or risk management action to control or reduce unacceptable risks, including the choosing of the actions, implementation, testing and monitoring to validate effectiveness
safety factor	An application factor applied for determining ecotoxicological criteria (thresholds or safe levels of pollutants in organisms)
slope factor	An indication of how powerful a chemical is in causing cancer expressed as the cancer risk per unit of dose (risk per mg/kg/day) (USEPA)
source	The hazardous substance/agent
soil allocation factor (SAF)	The proportion of TDSI that can be allocated to a site
soil concentration factor (C_S)	The unknown soil concentration value.
target	See *receptor*
tolerable daily intake (TDI)	An estimate of the average daily intake of a contaminant, expressed in terms of $\mu g\ d^{-1}$, that can be ingested over a lifetime without appreciable health risk. This is the UK terminology similar, but not identical in definition to, the US reference dose (RfD – a threshold dose for non-genotoxics and non-carcinogens)
total estimated daily intake (TEDI$_{SS}$)	The intake, or dose, of a contaminant from a site for all relevant pathways.
tolerable daily soil intake (TDSI)	The maximum intake of a substance that can be allocated to a contaminated soil.
uncertainty factor	Applied to the safety factor (when deriving RfDs and RfCs) to account for interspecies variability (USEPA)
upper confidence limit	A sample based estimate of the upper limit below which the true mean of the population will be located. A useful descriptor for testing whether the samples taken from a site are below or above a threshold value
volatilisation	The conversion of a chemical substance from a liquid or solid state to a gaseous or vapour state by the application of heat, by reducing pressure, or by a combination of these processes

Abbreviations

ADI	acceptable daily intake
ASTM	American Society for Testing and Materials
BI	background Intake
BRE	Building Research Establishment
CBI	Confederation of British Industry
CDM	Construction Design and Management Regulations 1994
CLRAM	Contaminated Land Risk Assessment Model
COC	UK Committee on Carcinogenicity in Food, Consumer Products and the Environment; some publications use this abbreviation for contaminant of concern
COSHH	Control of Substances Hazardous to Health Regulations 1999
CS	soil concentration factor
C$_{Scrit}$	critical soil concentration value
DETR	Department of the Environment, Transport and the Regions
DNAPL	Dense Non Aqueous Phase Liquid
DoE	Department of Environment (now DETR)
EA	Environment Agency
EC	effective concentration
EDI	estimated daily intake
EOD	explosive ordnance detection
EU	European Union
ERA	ecological risk assessment
HEAST	health effects assessment summary tables
HSE	Health and Safety Executive
GLC	Greater London Council
ICRCL	Interdepartmental Committee on the Redevelopment of Contaminated Land
IPC/IPPC	integrated pollution (prevention) control
IRIS	Integrated Risk Information System
LC	lethal concentration
LD	lethal dose
MAFF	Ministry of Agriculture, Fisheries and Food
MCL	maximum contaminant level
MDI	mean daily intake

MHSPE	Ministry of Housing, Spatial Planning and the Environment (Holland)
NAMAS	National Measurement Accreditation Scheme
NOEL	no observed effect level
NRA	National Rivers Authority (now EA)
NRPB	National Radiological Protection Board
OPC	ordinary Portland cement
OS	Ordnance Survey
PAH	polycyclic aromatic hydrocarbon
PPC	pollution prevention control
QA/QC	quality assurance/quality control
RAGS	Risk Assessment Guidance for Superfund
RBCA	risk-based corrective action
RfC	reference concentration (inhalation exposure)
RfD	reference dose (oral or dermal exposure)
RME	reasonable maximum exposure
SAF	soil allocation factor
SSSI	Site of Special Scientific Interest
TDI	tolerable daily intake
TDSI	tolerable daily soil intake
TEDISS	total estimated daily intake
TPH	total petroleum hydrocarbons
TSA	thaumasite sulphate attack
UKAS	United Kingdom Accreditation Service
USEPA	United States Environmental Protection Agency
WHO	World Health Organisation

1 Introduction

1.1 WHAT IS CONTAMINATED LAND?

"Contaminated land" generally refers to land that contains elevated concentrations of potentially hazardous substances. These concentrations may be present naturally, but more commonly, contamination is the legacy of the industrialisation of Britain over the past 200 years. The industries and processes that have been a feature of the landscape during this time have often caused the ground to become contaminated with the substances once handled at these sites. In many cases, the substances may be harmful to human health or the environment. More recently our understanding of the effects of these materials has developed. This increased knowledge has been accompanied by efforts to reduce or curtail their release into the environment and to manage and minimise their effects. The term "contaminated land" has a specific statutory meaning in the context of the regulatory regime under Part IIA of the Environmental Protection Act 1990. However, in this report the term is used in its wider, general sense.

1.2 WHY IS CONTAMINATED LAND IMPORTANT?

When land is contaminated it can affect human health, the environment (that is animals, livestock, plants, other organisms and micro-organisms, air, soil, the subsurface, groundwater and surface water), buildings and structures. The effects are not just toxicological; the value of land may be adversely affected by contamination. Contamination affects the uses to which land can be put. Put simply, there are many potential problems associated with contaminated sites that do not apply to uncontaminated sites.

1.3 WHY THEN CONSIDER CONTAMINATED SITES AT ALL?

The concept of urban decay, abandoned derelict sites within or on the edge of towns, is familiar. Central government policy is to encourage the reuse of derelict, or "brownfield", sites. The policy demands that 60 per cent of new housing be constructed on brownfield sites to relieve pressure to develop "greenfield" sites, to aid preservation of the countryside and to encourage urban regeneration. Some brownfield sites may be within locations that have become prime development sites. So although there might be extra costs associated with the redevelopment of these sites, this is often offset by an increase in the value of the land after the contamination issues have been addressed. There are thus economic reasons to consider selecting contaminated sites for redevelopment or ownership.

There are also statutory reasons why attention is given to contaminated land. Contaminated land is controlled by various legislative provisions, for example, those related to planning, waste management and water resources, and most recently Part IIA of the Environmental Protection Act 1990 and the Pollution Prevention Control Act 1999. Contaminated land is also associated with issues of liability in connection with civil law. In English law, for example, the principle of "caveat emptor" or "buyer beware" holds. Property transactions have always held an element of document search and interpretation and this may involve enquiries to identify contamination. A discussion of the legislative and regulatory regime appears in Appendix 2.

1.4 HOW MUCH CONTAMINATED LAND IS THERE?

The precise figure is not known. There is as yet no central record. Estimates vary from 50 000 ha to 200 000 ha (CBI, 1993). The Environment Agency (1999) estimates that some 300 000 ha of land across Britain may be affected either by industrial or "natural" contamination. Although not all these sites will pose immediate concerns, the Agency estimates that there may be between 5000 and 20 000 "problem sites".

1.5 WHY IS CONTAMINATED LAND RISK ASSESSMENT IMPORTANT?

Land that is contaminated can cause an unacceptable risk to human health and the environment. Sometimes it is not known whether contamination exists or whether the contamination that is present is likely to be hazardous to health or the environment. A need exists for a technique that identifies and considers the risks associated with such land, determines whether the risks are significant and whether action needs to be taken to reduce or control detrimental effects. This technique is called risk assessment.

A question frequently asked is "Is the contaminated site safe, are the levels of contamination acceptable?" The purpose of risk assessment is to provide an answer to the question:

> **"Is this site or area of land posing, or likely to pose, unacceptable risks to health or the environment?"**

1.6 THE PURPOSE OF THIS REPORT – WHO SHOULD BE INTERESTED IN THE GUIDANCE?

This report is directed at those who have to consider the risks posed to human health, the environment and the structure and fabric of buildings and infrastructure by land contamination (stakeholders, see Figure 1.1), who may then have to make decisions about what the contamination means to them based on their understanding of the assessment process.

Often the persons making the decisions are not technically qualified or specialists in the field of land contamination. These people, as well as those with more specialist knowledge of contaminated land risk assessment, need to understand the process involved so that they can make the most appropriate decisions. Because of the wide range of abilities and interest in detail that parties involved in risk assessment have, this guidance is designed to be of use to all those with an interest in risk assessment.

In general the report should be of interest to the following groups:

- those that specify and commission (or have an interest in) contaminated land risk assessment (*clients*)
- those that practice contaminated land risk assessment (*practitioners*); and those that regulate contaminated land (*regulators*).

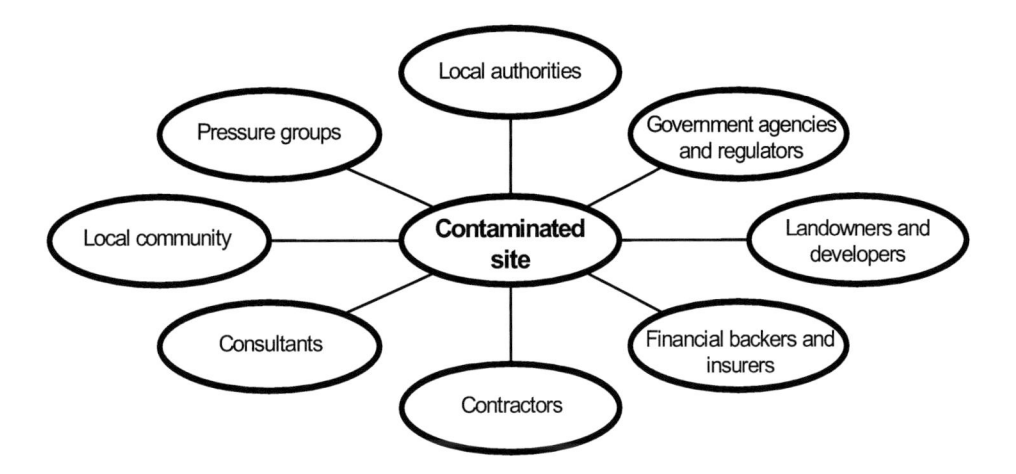

Figure 1.1 *Contaminated land stakeholders*

For each of these groups the objective is to raise awareness of the need for and the processes involved in carrying out risk assessment, and for those who practice, to raise levels of understanding of the procedures and technical guidance available in the UK. The report outlines good practice in the approach to assessment and provides technical and non-technical advice at each step with tips for good practice. The report also aims to raise awareness of how risk assessment fits in with issues such as contaminated land remediation and health and safety practices.

2 The basic framework for contaminated land risk assessment

2.1 INTRODUCTION

Chapter 1 described the key objective of contaminated land risk assessment, which is to determine whether land is posing or could pose unacceptable risks to human health, the environment, buildings and structures. Risk assessment is rarely carried out in isolation, however. Risk assessment of contaminated land normally forms part of an overall risk management strategy that involves making not only technical decisions, but also commercial, legal and financial decisions. Risk management also includes the development and evaluation of risk reduction measures, planning and maintenance of on-going control measures (such as monitoring) and communication of risk. The decision process may be driven by regulatory pressure, environmental standards, risk-cost-benefit analysis, public perception or "good practice".

This chapter positions risk assessment within the overall risk management process and describes the main stages of risk assessment.

2.2 THE RISK MANAGEMENT APPROACH

The risk management strategy for any site begins with the supposition that contamination could be present at a particular site and ends with conclusions made about the risks and decisions about how they might be controlled and managed. The risk management approach may involve activities beyond risk assessment and include the selection and implementation of remedial measures (the criteria for selection and implementation of remedial measures is beyond the scope of this report). Figure 2.1 illustrates the risk management process (including risk assessment and subsequent activities).

The motivation for making risk-based decisions may be created by the regulatory regime (Chapter 1) or may be driven by practical considerations, such as concerns about the health of the users of a particular piece of land (where this concern is not legislation driven). It may also be driven by the opportunity to reduce costs of subsequent site remediation through careful use of risk assessment techniques.

2.3 THE RISK-BASED APPROACH IN THE UK

Section 1.3 introduced some key legislation controlling contaminated land. Most recently, Part IIA of the Environmental Protection Act 1990 introduced a new statutory regime for the control and treatment of existing contamination (see Appendix 2.1.3). This is based on the "suitable for use" approach. This approach requires some sort of remedial action where it is shown that, at a particular site, contamination is causing (or potentially causing) unacceptable risks in relation the current or intended uses of the site (not requiring planning permission). The "suitable for use" approach involves managing the risks posed by contaminated land by making risk-based decisions. Risk-based decisions are a feature of much of the legislative regime as discussed in Appendix 2.

The risk-based approach to the assessment of sites is founded on the "contaminant–pathway–receptor" relationship, using the following definitions.

Contaminant (sometimes known as the "source") – the hazardous substance/agent. In many cases this will be a potentially hazardous chemical present on or in the ground at concentrations that are deemed potentially hazardous

Receptor – the entity (eg human, animal, water, vegetation, building services) that is vulnerable to the adverse effects of the hazardous substance or agent. May also be called the "target".

Pathway – the means by which a hazardous substance or agent comes into contact with, or otherwise affects a receptor.

For a risk to exist there must be a contaminant capable of causing harm, a receptor sensitive to that contaminant and a pathway linking them. If there is no link between the contaminant and receptor then there will be no unacceptable risk to that receptor posed by the contaminant.

The contaminant–pathway–receptor relationship is often known as a "pollutant linkage". On an individual site there may be more than one such pollutant linkage, and each of these require individual assessment. An effective risk assessment aims to properly identify all the pollutant linkages.

In broad terms, risk assessments consider the following types of receptor:

- **humans**
- **environment, ie water (ground and surface waters), flora and fauna**
- **structure and fabric of buildings and infrastructure**

For the purposes of the Part IIA regime, the specific receptors within these groups are identified (Table 2.1).

2.4 THE MAIN STAGES OF RISK ASSESSMENT

The risk assessment process can be broken down into four main stages. These are:

- hazard identification
- hazard assessment
- risk estimation
- risk evaluation.

Communication of risk may occur at any stage. The key activities involved at each stage are as follows:

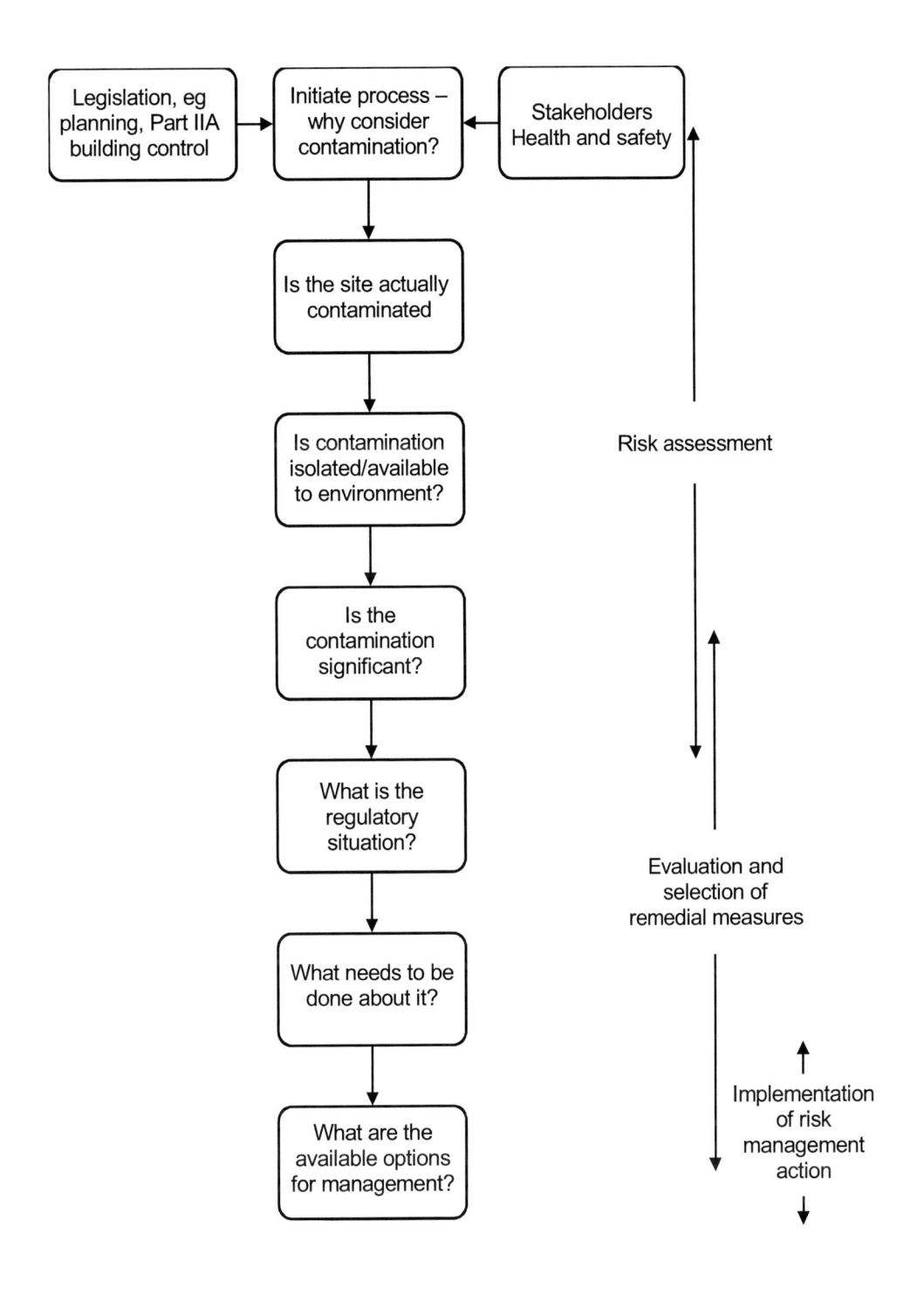

Figure 2.1 *The risk management process*

Table 2.1 *Receptors, for the purposes of the Environmental Protection Act 1990**

Receptor	Details
Human beings	
Ecological systems or living organisms forming part of a system within a location which is (shown opposite):	• an area notified as an Site of Special Scientific Interest under Section 28 of the Wildlife and Countryside Act 1981 • any land declared a National Nature Reserve under Section 35 of that Act • any area designated as a Marine Nature Reserve under Section 36 of that Act • an Area of Special Protection for Birds, established under Section 3 of that Act • any European site within the meaning of Regulation 10 of the Conservation (Natural Habitats etc) Regulations 1994 (ie Special Areas of Conservation and Special Protection Areas) • any habitat or site afforded policy protection under Paragraph 13 of Planning Policy Guidance Note 9 on nature conservation (ie candidate Special Areas of Conservation, potential Special Protection Areas and listed RAMSAR sites) • any nature reserve established under Section 21 of the National Parks and Access to the Countryside Act 1949
Property in the form of:	• crops, including timber • produce grown domestically, or on allotments, for consumption • livestock and other owned or domesticated animals • wild animals that are the subject of shooting or fishing rights
Property in the form of buildings	• "building" has the meaning given in Section 336(1) of the Town and Country Planning Act 1990 (ie it includes "any structure or erection, and any part of a building.... but does not include plant of machinery comprised in a building") • ancient monuments
Controlled waters	• surface waters, groundwater

*Ref: DETR (2000), *Circular 02/2000: Environmental Protection Act 1990: Part IIA Contaminated Land*, HMSO, London.

a) Hazard identification

Firstly the site is defined and boundaries identified. The stakeholders are identified and the assessment objectives finalised. Potential contaminants and receptors (targets) are identified and potential routes by which the two may be linked are determined. A preliminary (conceptual) contaminant–pathway–receptor risk assessment model is proposed (see Chapter 4).

b) Hazard assessment

At this stage the hazards, receptors and linkages between them identified in (a) above are assembled and reviewed to see whether these are realistic and representative of the expected conditions at the site. This provides a preliminary indication of potential short-

or long-term risk to the environment. Further data may be required at this point before proceeding further. Consideration of the data collected may result in refinement of the conceptual model.

c) Risk estimation

A site investigation including measurements, sampling and analysis may be required at this stage. The conceptual contaminant–pathway–receptor model proposed in (a) above is refined by reference to the information collected in (a) and (b) and the site investigation. The potential for and frequency of occurrence of the risk is estimated for each pollutant linkage and estimates are made of the possible effects on each receptor.

d) Risk evaluation

Decisions are made about the significance of the risk and the measures that must be taken in order to reduce the risks to acceptable levels. Final communication of the outcomes of the assessment process is made to all interested parties.

The above process can be summarised – see Figure 2.2.

Hazard identification	⟶	What are the possible problems?
Hazard assessment	⟶	How big might these problems be?
Risk estimation	⟶	What will be their effects?
Risk evaluation	⟶	Do they matter?

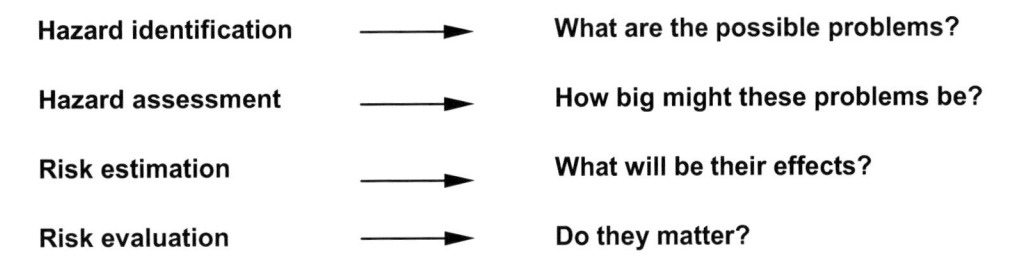

Figure 2.2 *Summary of the risk assessment process*

Figure 2.2 indicates the activities involved in the risk assessment process. The practical enactment of these normally requires the tasks involved to be divided into distinct phases as follows:

Phase 1 risk assessment – involving document collection, site visit and conceptual model preparation (hazard identification and hazard assessment). (See Chapter 4).

Phase 2 risk assessment – this builds on Phase 1 assessment and involves site investigation, measurements, sampling, testing and evaluation involving data analysis, risk evaluation and communication (risk estimation and risk evaluation). (See Chapters 5 and 6).

Sometimes Phase 1 is further divided into two sub-phases, Phase 1a and Phase 1b, as follows.

Phase 1a risk assessment – incorporates the conceptual stage of hazard identification. It involves obtaining data on the characteristics of a site in order to develop the conceptual model – hazard identification

Phase 1b risk assessment – builds on Phase 1a risk assessment findings and incorporates the conceptual stage of hazard assessment. It involves refining the model and making preliminary decisions about the risks posed to human health, the environment, buildings and structures, with collection of more information if needed.

The main activities involved in risk assessment are shown in Figure 2.3.

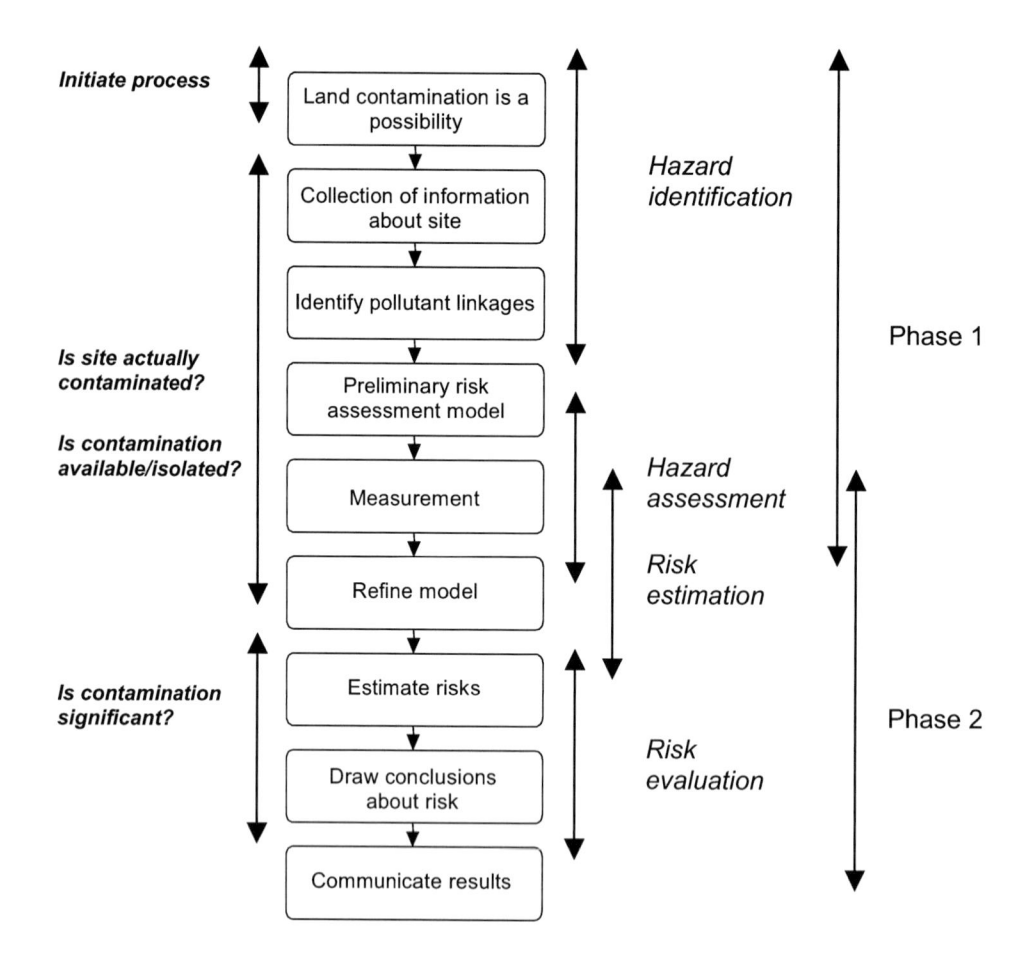

Figure 2.3 *Process for risk assessment*

Throughout the risk assessment process the assessor should aim to obtain sufficient information about pollutant linkages as to be confident about decisions made concerning risks. Although the assessment as described above is a sequential process, not every assessment will involve each of the above activities. Completion of each activity may not be necessary depending on the objectives of the risk management process. The process is iterative, so many of the activities are repeated until sufficient information is obtained to allow the assessment to move forward. It is important that the decision-making context should be decided at the outset of the assessment and a dialogue established with all the stakeholders. The determination of risk management objectives is described in the next chapter.

2.5 FURTHER READING

DETR (2000)
Circular 02/2000: Environmental Protection Act 1990: Part IIA Contaminated Land.
HMSO, London
A useful discussion of the new contaminated land regime and how it interacts with other regimes.

HARRIS M R and DENNER J (1997)
"UK Government Policy and Controls". In: Hester, R E and Harrison, R M (eds):
Contaminated Land and its reclamation, pp 25–46
Thomas Telford, London
A useful background description of the form and development of UK government policy on contaminated land.

3 Defining risk assessment objectives

3.1 INTRODUCTION

The risk assessment should begin by setting the management context within which the assessment is to be carried out. The objectives and key requirements for the work to be undertaken need to be decided at this time. It is important at the outset that the context and requirement of the assessment are clearly understood by all parties involved as part of a clear, transparent process enabling all stakeholders to follow and participate at all the stages of the assessment.

3.2 DECIDING THE CONTEXT

The various stakeholders (identified in Chapter 1) involved in a particular potentially contaminated site will have different motivations and expectations of the assessment process and its outcome. The direction and scope of the risk assessment for a particular site will depend on the particular stakeholders involved, the regulatory situation, the plans for the site and the funds and time available.

Risks associated with the site are assessed for a variety of reasons. Risk assessment is applicable in a variety of commercial circumstances (such as supporting pre-purchase funding decisions) and under different legal regimes (such as planning or Part IIA of the Environmental Protection Act 1990). The following are typical objectives of risk management:

- simple qualitative assessment of risk prior to a site visit. This is a health and safety assessment undertaken to ensure the safety of the assessor during their visit to site
- qualitative assessment for the prioritisation of several sites. Where there are many sites to assess, risk-based screening will determine the most urgent cases for review
- determination of the need for remedial action
- assistance in the valuation of the site
- assessment for the purposes of Part IIA of the Environmental Protection Act 1990
- demonstration that risk has been reduced following remedial actions taken at a site (validation of remedial actions).

The assessor should establish at the outset the context within which the assessment is being carried out and the likely risk management decisions that will be made based on the findings of the assessment. Some examples are provided in Box 3.1 of the kind of risk management decisions that will be made on the basis of the assessment.

Box 3.1 *Examples of risk management objectives*

---

1. A manufacturing company owns a considerable number of sites. The property of this company is acquired by a property holding company, whose environmental policy requires it to investigate and assess all potentially contaminated sites within its holding. A sum of money has been set aside by the holding company for investigation and remediation of any sites that are deemed to be causing harm to human health or the environment. The holding company needs to have a reasonable, workable and objective scheme to assess each site within its holding. It needs to prioritise the sites in terms of those posing immediate risk of harm (necessitating urgent action) and those not requiring immediate action but where remediation in the long term may be desirable. This is in order to allow decisions to be made on funding of investigations and remediation.

In this example, the funding was approved for Phase 1 assessments, with the funding for site investigations being directed at the time to those sites deemed to be posing imminent risk of harm.

---

2. A developer wishes to construct a warehouse complex and plans to position the complex within the site, balancing necessary car parking and landscaped areas with those covered by buildings, in accordance with outline planing consent for the site. The developer uses risk assessment techniques to minimise the disturbance of ground, by modifying the relative position of buildings, parking and landscaped areas. The result is a saving on the disposal/treatment costs of soil excavated during the subsequent construction works.

In this example a relatively high degree of technical certainty is required, because it is important to fully identify, characterise and delineate the contamination with the objective of allowing earthworks quantities to be estimated as accurately as possible.

---

3. A local authority is carrying out its duties under Part IIA of the Environmental Protection Act 1990. A site adjacent to a river is thought originally to have been a valley infilled with domestic and industrial wastes. A risk assessment is conducted with the aim of establishing the likely presence of contaminating substances, their extent, possible pathways for these substances to migrate from the site and the relative location of sensitive environmental features.

The likely presence of highly contaminated materials is established, and pathways for the migration of contamination off site are identified, by Phase 1 studies, both of which are confirmed by site investigation. An estimate of risks posed to humans and the environment is prepared and as a result the decision is taken to require remedial action to curtail lateral movement of contaminants.

For this site the risk assessment established potential pollutant linkages, determined the nature and extent by site investigation and estimated the probability that "significant harm" might occur.

3.3 THE CONSEQUENCES OF RISK ASSESSMENT

When the scope of risk assessment is being decided some thought is usually given to how the findings of the assessment can be applied to risk-based decision-making. Consider the example in Box 3.2.

Box 3.2 *Example of a decision based on risk assessment*

> A potential purchaser is considering whether to acquire a site that has a history of industrial use but for which there is no available data relating to the potential for the presence of contamination. The objective of the risk assessment is to provide sufficient data for the purchaser to make an informed decision concerning their potential liabilities on acquiring the site.
>
> A Phase I assessment is conducted and indicates the potential presence of considerable contamination. The purchaser decides not to acquire the site. This is on the basis that the likely reduction in the site's value as negotiated with the present owner (because of the presence of contamination) is insufficient to offset the increased time and expense needed to remediate the site to an acceptable standard.

3.4 AGREEMENT OF THE SCOPE OF THE ASSESSMENT

Agreement of the scope of the assessment at the start of the management process is essential to ensure that all parties know what is required of them and what they should expect from the assessment.

At the beginning of the assessment the assessor needs to agree with all relevant interested parties the detailed scope and constraints. The assessor needs to know the context within which the assessment is being carried out and should determine how the risk assessment will be used to make risk-based decisions. Items to be agreed will be both technical and non-technical (Table 3.1). They will need to agree the budget available, the time period within which the assessment will be completed, the acceptable degree of technical uncertainty (primarily dependent on available time and budget) and requirements for confidentiality. This last element is particularly important since a purchaser may not wish to reveal their intention to acquire property at the early stage at which the assessment may be conducted. The need for confidentiality may limit the opportunity for enquiries to public bodies such as the Environment Agency, and the consequences of this (a possible gap in the information on the site available to the assessor) should be understood. In contrast, the need to communicate risks, timing and constraints on risk communication should be agreed (see Chapter 7).

Table 3.1 *Examples of technical and non- technical considerations*

Technical considerations	Non-technical considerations
Degree of technical confidence	Requirements for confidentiality and communication
Acceptability criteria	Identification of interested parties
The area of ground to be assessed – the extent of site to be investigated	Budget
Site boundaries and access (contamination is no respecter of boundaries)	
Scope for sampling and analysis (also dependant on budget)	Time available
Requirements for long-term monitoring	Regulatory regime

3.5 PROCEEDING WITH THE ASSESSMENT

The procedure for risk assessment will follow the four main stages outlined within Section 2.4 and is an iterative process.

```
Hazard identification
        |
Hazard assessment
        |
Risk estimation
        |
Risk evaluation
```

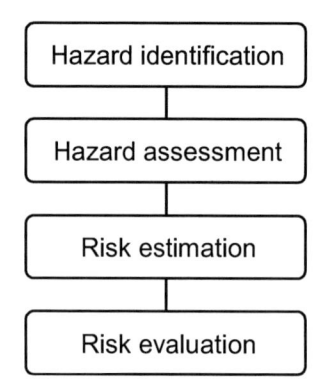

Figure 3.1 *The sequential process for risk assessment*

In most cases the assessment begins with identifying the hazards and concludes with the evaluating and communicating the significance of the risks that have been estimated. Not all of these activities will be carried out all of the time. Some assessments will cease after Phase 1, for example when a potential purchaser understands the risks sufficiently to be able to make a decision. Where a landowner has a portfolio of several sites but only limited funds to effect works to control contamination, the Phase 1 assessment may culminate in risk ranking. This requires a systematic approach to deciding the priority to give to action on a site that may be contaminated. The ranking procedure chosen, in this context, will allow sites to be ranked according to the potential hazards posed. Phase 2 risk assessments may follow on from the Phase 1 assessment; however, circumstances are unlikely to arise where Phase 2 assessments are carried out in the absence of Phase 1 information. Where Phase 1 data has not been acquired, the Phase 2 investigation may be unfocussed, since potential pollutant linkages will not have been identified. The Phase 2 investigation would effectively be examining pollutant linkages, which may not be reasonable in the light of the site setting past and present, and there is a significant possibility that areas of contamination may be overlooked.

Box 3.3 provides some tips for good practice.

Box 3.3 *Tips for good practice*

> On many occasions a risk assessment will extend only as far as Phase 1 assessments. As a result, the absence of numerical data indicating actual concentrations of contaminants will mean that the level of technical uncertainty will be quite high. The assessor should consider whether the balance of commercial risks associated with this uncertainty are acceptable.
>
> It is advisable at the outset to consider how the results of the risk assessment would be relevant to all of the interested parties. If required the scope of the assessment can be modified to address this issue.

4 Phase 1 risk assessment – gathering information

4.1 WHY IS THE INFORMATION-GATHERING STAGE NECESSARY?

The first practical step of the risk assessment is to gather information in order to establish whether a site may be contaminated. If contamination is suspected, the study should seek to establish the presence, nature and likely extent of contamination, the pathways and receptors and to identify any areas where further site-specific data is needed, such as the need for detailed physical site investigations. This Phase 1 investigation includes **hazard identification** and **hazard assessment** and culminates in the development and refinement of the preliminary conceptual model.

Some pertinent questions are asked:

- what is the history of the site and its environs, and what is its environmental setting? This establishes a preliminary indication of the likelihood of finding contamination
- what is the site's intended use, or, if no development is proposed at the site, what is its current use? In contaminated land risk assessment, the significance of contaminants at the site is related to the ongoing or planned use of the site.

The Phase 1 study represents the assessor's first look at the site and first opportunity to view factual data on the site. Throughout the risk assessment process the assessor's aim is to gather sufficient information on the site (or sites) to increase understanding of the pollutant linkages and ultimately to enable confident judgements to be made as to the significance of the risks. As more information becomes available the uncertainties associated with the assessment are reduced.

The Phase 1 objectives are:

- to check the likelihood of the presence of contamination that may affect the suitability of the site for a specific current or future use
- to indicate its nature and effect, pathways and receptors
- to identify special precautions and procedures to be taken during operations on the site itself (such as site visits and investigations)
- to provide information from which an effective site investigation could be designed should this be required.

Failure to identify a significant hazard (and pathway and receptor) at this stage in the process could result in risks that remain unidentified and thus uncontrolled.

4.2 HOW IS THE INFORMATION GATHERED?

The Phase 1 assessment relies on document gathering, review and interpretation with parallel consultation with interested parties, and a preliminary visit to the site. The information should be gathered sequentially to characterise the site.

Several distinct steps are involved in Phase 1 assessment.

Step 1 Define the boundaries of the site (extent of study area)

Step 2 Obtain historical and other information (including consultation). Preliminary visit to site (hazard identification)

Step 3 Collate information (hazard identification)

Step 4 Identify plausible contaminants, pathways and receptors, and the linkages between them (pollution linkages) (hazard identification)

Step 5 Develop the conceptual contaminant-pathway-receptor model from Step 4. Review plausibility and refine with additional data if required (hazard assessment)

Step 6 Decision-making – draw conclusions and decide whether risk assessment objectives have been fulfilled.

4.3 THE STEPS OF PHASE 1 ASSESSMENT

4.3.1 Step 1 – define boundaries

The first step is normally to agree the physical limitations of the study including the actual areas of ground that are to be assessed. The boundaries of the study site may be indistinct and should be defined at the start. It is good practice to define the boundaries clearly on an Ordnance Survey base at an appropriate scale. As the study progresses it is common to discover that the boundaries have changed over the years and that a particular contaminating activity originally transcended the present-day boundaries. It should be recognised that the effects of contamination could be apparent off site; alternatively contamination originating from an off-site source could impact upon the study site. The Phase 1 study should be flexible enough to allow for collection of information beyond the site boundaries as appropriate.

4.3.2 Steps 2 and 3 – obtain and collate information

At this stage, information is gathered that establishes the likelihood that a site is contaminated. Usually the study starts by determining the history of the site and adjoining areas, paying particular attention to identifying and determining the type of industrial processes or other activities that might have been contaminative. Box 4.1 describes the information that is normally required. There are numerous sources of information. Several publications give detailed advice as to sources of documentation and how these should be interpreted (see DETR, 1994 (CLR 3)).

Box 4.1 *Information that is normally required for Phase 1 assessments*

History of the site, previous boundaries, uses and users:

- past and current editions of Ordnance Survey maps
- past editions of town and local plans
- past editions of trade directories
- various generations of aerial photographs
- waste disposal and local planning authority and other statutory records
- newspaper archives and other documentary information
- anecdotal evidence from former employees and local residents (if available)
- local historians and history societies
- documentary evidence from existing or previous occupiers/owners (private and industrial archives)
- depending on the locality, other records such as coal and other mining records should be consulted.

Activities adjacent to the site:

- landfills
- contaminative processes
- nearby watercourses, water extraction/discharge
- Sites of Special Scientific Interest, other designated sites and sensitive environmental features.

Presence of unrelated features such as unauthorised tipping, services

Geology, hydrogeology and surface water courses and rising groundwater

Earlier risk assessments:

- other risk assessment reports
- sampling and testing data relating to the site
- information relating to the local environment (for example, reports describing mining groundwater).

What information sources should be consulted and how useful are they?

Table 4.1 lists some of the most commonly consulted sources of information for Phase 1 assessments and their relative usefulness. The list is not exhaustive and the documents that are most relevant will be very specific to the site. There are specialist commercial organisations who provide summaries of archived data and who will carry out searches of these sources on the assessor's behalf.

Talking to third parties

It will often be necessary to consult with interested parties as a component of the Phase 1 study, where these have information of use to the study. Table 4.1 details some of the information that is available from third parties and its relative usefulness. The assessor should allow sufficient time for information from these sources to be received.

The site reconnaissance

The site walkover is generally the investigator's first look at the site. Before entering the site the assessor should consider whether any particular health and safety precautions are necessary, bearing in mind the information about the site that has already been gathered by this time. During the visit, the assessor relates any physical features seen, such as old foundations, voids etc. to the historic plans and other information. They look at the practicalities of entering the site with invasive site investigation equipment (if appropriate) – access arrangements, overhead cables, mounds of materials. These are the kind of facts that are unlikely to be available from any documents. The presence of undocumented hazards, such as fly-tipped wastes, should be noted.

Previous site investigations and risk assessments

Sometimes some technical investigation data may exist for the site. This may have arisen as a result of former planning applications, monitoring initiatives by the Environment Agency or its predecessors, or as a response to known environmental problems at the site. This data should be collated; if compatible, it should be combined with data collected during subsequent phases of risk assessment. The assessor should be aware that this data may have been gathered in pursuit of different objectives. In addition the level of technical uncertainty and quality may make this data unsuitable for incorporation into the current study. For example, the available data might include chemical test data for soil samples. However the method by which the samples were analysed may not be appropriate for the current study (for example, the lower limit of analytical detection may be above the assessment concentration value deemed significant – see Chapter 6).

All the data should then be collated and presented sequentially to result in a detailed history of the site, indicating textually and diagrammatically where potentially contaminated activities have been located within the site.

Table 4.1 *Some key sources of information and their usefulness*

Information source	What it contains	Relative usefulness* 1	2	3
Maps: Ordnance survey Series 6 inch and 1:10 000	Series of maps covering all Great Britain	✔	✔	
Maps: Ordnance survey Series 25 inch (1:2500)	Earliest editions from 1854, all complete by 1893. Covers all Great Britain except urban areas covered by 1:1250 series since 1945 and areas of mountain and moorland covered by 1:10 000 series	✔		
Maps: Ordnance Survey 50 inch	Since 1945. Covers only urban areas. Useful for densely populated areas.	✔		
Maps: town maps, 1:500	For towns and cities. Most mapped only once, 1860s to 1890s		✔	
Maps: other Ordnance Survey based maps	Mostly small-scale (eg 1:50 000). Sometimes show details absent on other OS maps such as military installations.		✔	
Maps: tithe maps early to mid-19th century	Useful pre-OS information	✔		
Maps: enclosure plans, settlement plans, early county maps, parish and world maps, local town plans	Useful pre-OS information (beware of accuracy)	✔		
Maps: plans deposited in relation to statutory undertakings (such as railway plans in connection with railway Acts of Parliament)	Detailed information usually with limited lateral coverage		✔	
Maps: Goad insurance plans	Insurance plans for individual/blocks of buildings in town centre. Very detailed information.		✔	
Maps: geological maps produced by the British Geological Survey	Details of local geology – solid and drift maps and other issues, eg radon risk.	✔		
Maps: Environment Agency (NRA) Groundwater Protection Zone mapping	Information on groundwater vulnerability	✔		
Registers/archives: waste management licences	Held by the Environment Agency, lists all current and sometimes past, sites licensed under Environmental Protection Act 1990 (formerly licensed by the Control of Pollution Act 1974).	✔		
Registers/archives: statutory planning registers	Records of decisions made under the Town and Country Planning Act 1990 by local planning authorities	✔		
Registers/archives: records of premises regulated under the Radioactive Substances Act 1960 (later 1993)	Information on sites where radioactive substances are kept or used	✔		

Table 4.1 (cont'd)   *Some key sources of information and their usefulness*

Information source	What it contains	Relative usefulness* 1	2	3
Registers/archives: records of statutory undertakers	Information on the location of statutory undertakers' routes (electricity, gas, water etc.). Particularly important if ground disturbance is planned		✔	
Registers/archives: records of abstraction licences, instances of flooding and discharge consents held by the Environment Agency	Information on water uptake and discharge to ground surface and water courses, and areas of land vulnerable to flooding	✔		
Registers/archives: trade and street directories	Information on activities carried out at individual premises		✔	
Registers/archives: private and industrial archives	Very detailed information on previous activity at site, relating to location of plant, storage of materials etc	✔	✔	✔
Registers/archives: records of burial sites of diseased animals	Information on the burial sites of diseased animals, held by MAFF, water companies and local authorities (in the case of anthrax).	✔		
Registers/archives: coal authority records	Records of former mine workings	✔	✔	
Other sources: DoE industry profiles	Information on industrial and commercial processes, published by the Department of the Environment (now DETR)	✔		
Part IIA registers and Section 143 registers	Records of regulatory action in connection with Part IIA of the Environmental Protection Act 1990. Any registers that may have been compiled with the now abandoned Section 143 of the Environmental Protection Act 1990.	✔	✔	

* Notes on usefulness
1 Essential source, to be consulted on each occasion. All available editions to be examined.
2 Useful in many situations to provide additional information to confirm information available from Category 1 sources
3 Useful background information when other sources incomplete or unreliable.

4.3.3 Step 4 – pollution linkages

The data collated during Steps 2 and 3 should be reviewed to determine whether hazardous substances are potentially present at site. The nature of these substances, *contaminants*, should be identified. These hazardous substances will normally be the types of contaminant that would be associated with the activities carried out at the site. The Department of the Environment (now DETR) has published Industry Profiles (see Appendix 1) that describe typical contaminants associated with common industrial processes, and this may be used as a primary source of information. The presence of other substances may be suspected from the site walkover. Substances likely to be present naturally at elevated concentrations should be considered (such as radon gas, a natural decomposition product of the decay of radium and uranium in bedrock). Table 4.2 gives a description of the hazardous effects of certain classes of substances.

The potential features that may be affected by these substances, *receptors*, should then be identified. Some thought should be given to how each receptor might be exposed to the contaminants (that is, might come into contact with the contaminant) and the adverse effects that might be experienced.

Table 4.2 *Overview of selected contaminants and associated hazards*

Contaminant type	Possible hazards
Metals	Potential for toxic effects, possible carcinogenic effects in humans. Some metals cause toxic effects in animals
Phytotoxic metals	Toxicity to plants, depending on species tolerance
Anions	Affect ecosystems (eg eutrophication and salinity, damage to concrete (sulphate and chloride)
Hydrocarbons	Many are carcinogenic to humans, some are toxic to humans. May be toxic to animals and plants. Inhibition of normal concrete curing process
Corrosive and aggressive materials	Acute risks to humans by direct contact. Damage to buildings and structures
Asbestos	Carcinogenic risk to humans and animals
Explosives and munitions	Acute explosion and flammability risks, also toxicity to humans. Damage to buildings and structures
Radioactive substances	Human and animal toxicity, inhibition of plant growth.
Leachates	Likely to be toxic to aquatic life, aggressive to materials and structures. May enter groundwater
Soil gases	Fire, explosive and asphyxiant risk to humans, inhibition of plant growth
Pathogens	Infectious disease hazards to human and animals

When all the contaminants and receptors have been identified, the possibility that there might be a link between them should be assessed (*Pathways*). Where contaminants, receptors and a likely pathway between them are identified, these are known as pollutant linkages (Chapter 2). It should be remembered that if development is proposed this could alter the pathways for contaminant movement, and this should be allowed for when potential pathways are being identified.

An example illustrating this hazard identification stage and identification of pollutant linkages is given in Box 4.2. This example illustrates the importance of identifying potential receptors allowing for the site in both its current and likely future state, as far as this is reasonably foreseeable.

Box 4.2 *Example illustrating the determination of pollutant linkages*

> It is intended to construct a retail warehouse on the fringes of a town. The site for the warehouse is currently unoccupied. The Phase 1 studies, including the document research and enquiries, establish that the land immediately adjoining the site to the west is a former sand and gravel quarry infilled with domestic wastes. A study of geological maps and some data available from old borehole records at the site indicates the geology consists of made ground/topsoil over around 6 m thickness of sand and gravel overlying a considerable thickness (30 m) of dense unfissured clay. An assessment of the hydrogeological regime suggests groundwater within the sand and gravel and indicates general groundwater movement in a south-easterly direction.
>
> Consideration has been given to pollutant linkages. The domestic waste within the quarry is around 30 years old and would be expected to be producing landfill gases as a result of decomposition. The main gases produced would be methane (a hazard since at the appropriate concentrations it is explosive and asphyxiant) and carbon dioxide (asphyxiant). Decomposition would lead to the production of leachate, liquid containing dissolved components of the waste and its products (some of which might be harmful to human health, the environment and even buildings). These would constitute potential ***contaminants***. The research and site visit give no indication of the presence of any measures to control the products of decomposition, so it is anticipated that these could migrate via two main ***pathways*** from the infilled quarry into the development site:
>
> - **Pathway 1**: via the relatively permeable sand and gravel stratum
> - **Pathway 2**: via the shallow groundwater within the sand and gravel
>
> Several potential ***receptors*** could be brought into contact with the migrating gas and leachate, both with the site in its present state and in relation to its intended use. As the site is currently undeveloped, the groundwater itself is one receptor. The proposal to develop the site could bring site investigation and construction staff into contact with contaminated soil gases and leachates and, in the longer term, future users at the site might potentially be at risk from soil gases collecting under or within the building. There are thus several ***pollutant linkages***:
>
> - Pollutant linkage 1 – entry of leachate into groundwater
> - Pollutant linkage 2 – entry of soil gases into the development site adversely affecting plant growth
> - Pollutant linkage 3 – entry of soil gases and contaminated leachate into the development site posing risks to construction site and future users of the site.

4.3.4 Step 5 – developing the conceptual contaminant-pathway-receptor model

The example in Step 4 above proposes several pollutant linkages. All the potential pollution linkages should be assessed individually and together. The objective is to identify all the potential hazards to health, environment or infrastructure. This can be summarised as follows:

- hazards to human health
- hazards to the water environment
- hazards to flora and fauna
- hazards to the fabric of buildings, structures and services.

These are discussed in Chapter 6. A checklist is recommended to ensure that all relevant hazards have been considered.

Possible contaminants (hazard identification), receptors and pathways

Are any of the following known or suspected:

- manufacturing processes leaving residues (raw materials, products, by-products, wastes)
- spillages, dumping, fly-tipped materials, fires
- waste tips (licensed and unlicensed)
- soil gases
- as above on adjoining sites.

Are any the following potential receptors present? (Note, for the purposes of Part IIA, receptors are specifically defined within guidance associated with the regulations):

- humans on or in the vicinity of the site
- sensitive environmental features, eg plants, watercourses
- groundwater – for example, is there groundwater abstraction near the site or is the site in an aquifer protection zone?
- building infrastructure or services.

Are any of the following pathways suspected?

- uptake by receptors via ingestion
- uptake by receptors via inhalation
- uptake by receptors via eye contact
- uptake by receptors via absorption through skin
- uptake by receptors via consuming animals or plants
- uptake by vegetation
- uptake by receptors via consuming water
- migration via permeable strata (eg ground, services or building materials).

The conceptual model can be developed on the basis of the information collected. The conceptual model is an illustration of the relationship between the identified contaminants, pathways and targets. It is a means of representing the characteristics of the site allowing the pollutant linkages to be clearly reviewed for their reasonableness. The model may be expressed in a tabular, matrix or pictorial format (see Table 4.3). The model can be updated to reflect new information about the site as this is discovered. The model should identify:

- each receptor
- each pathway by which the receptor becomes exposed
- each substance thought to be present at site.

There should be some indication in the model of the likelihood of the pollution linkage being completed (in the example in Table 4.3, the likelihood of a probable linkage is indicated by). The model should also assess any reasonable likelihood that pathways might change with time, for example were groundwater to rise and mobilise contamination. The model should take into account the effects of construction or remedial activities, which albeit temporary, may create new pathways for contaminant migration. This is discussed in Box 4.3.

An example of a tabular model (based on the example in Box 4.2) is shown in Table 4.3.

Table 4.3 *Example of a preliminary conceptual model in tabular form*

Receptor	Pathway	Substances Leachate	Soil gas
Water environment	leachate enters groundwater within site		
Human health	soil gases and leachates enter confined spaces during construction (short-term)		
Human health	soil gases enter building (long-term)		
Human health	ingestion/inhalation of contaminated leachate by site users (long-term)	X	
Building materials	contact with contaminated soil		

– plausible pollutant linkage is likely
X – plausible pollutant linkage is unlikely

Box 4.3 *Risk assessment in relation to development and construction activities*

A risk assessment in connection with proposals for redevelopment or construction activities should consider how new pathways might be created for contaminants to reach receptors. Any form of ground disturbance may mobilise contaminants into the atmosphere, to groundwater or to surface water, or may expose contamination that might become a hazard to users or workers at a site. A developer may undertake an assessment of this nature as part of the risk assessment of the site, as part of the COSHH (Control of Substances Hazardous to Health Regulations) assessment that is required before construction starts and as part of the health and safety requirements of the Construction (Design and Management) Regulations. It is advised that this assessment is carried out as early as possible to enable risk management solutions to be developed that allow identified potential risks to be investigated and controlled.

A preliminary assessment of the likelihood that contamination will be mobilised by construction activities may be carried out during Phase 1. The assessment will consider the possible effects of the construction:

- nuisance issues – the effects of dusts, odours and other emissions on neighbours
- the opportunity for contaminated runoff to affect adjoining land and watercourses
- effects on farmers and commercial growers. Plant growth may be affected by dusts especially those that are chemically active, such as limestone
- constraints on construction activities. The presence of contaminants may delay construction while being removed and may affect the phasing of projects
- effects on wildlife and natural features
- additional costs incurred in removing contaminants.

The assessment may require measurements to be made of the current condition of the site (see Chapter 5). The assessment should address the following key questions:

- Will the development expose new or mobilise existing contaminants?

  Excavations into contaminated land may release vapour or dust into the atmosphere, potentially affecting neighbouring properties as well as site workers. Stockpiles of soil awaiting disposal, treatment or reuse may generate runoff, potentially affecting surface or groundwater resources, or require disposal to foul sewer.

- Will the development create new sources of contamination?

  Potential new contaminants can be introduced during construction such as from oil/fuel storage tanks.

- Will new pathways be created during construction activities?

  New pathways can be created during construction such as surface water runoff into drainage systems, which may affect local watercourses and groundwater. Dust generation along haul roads or during processing activities such as screening can generate dusts, which may be contaminated.

4.3.5 Step 6 – drawing conclusions and fulfilling objectives

At this point the adequacy of the data collected should be assessed. Uncertainties and gaps in the data should be identified and, where necessary and possible, rectified. The assessor should be aware that risk assessments are subject to a degree of uncertainty throughout the assessment process. Uncertainties arise when there are gaps in the information and can be reduced by revisiting the historical and other data sources consulted earlier in the assessment. Since it is not possible to eliminate every uncertainty beyond all doubt, it is usual to make assumptions where there are information gaps. For example, the assessor may not have been able to discover any details about the site for a 20-year period in its history. They may choose to assume that activities at the site did not differ significantly from those recorded during the periods immediately before and after the unknown interval, perhaps concluding that no additional contaminating activities took place over that period. The assessor should determine whether the assumptions made about the site could affect the subsequent judgements about risks.

Depending on the objectives of the assessment this may be the end-point of the risk assessment. The following outcomes from Phase 1 hazard identification and assessment are possible:

1) There is no evidence for any potential contaminant–pathway–receptor linkages, so no further action is required.

2) There is evidence for completed contaminant–pathway–receptor linkages, but there is no legal or other obligation to take further action. For example, the indication that contamination is likely to be present may be enough to allow a decision to be made about whether or not to proceed with the purchase of a property (although the decision may be taken at a low level of technical confidence).

3) Where a Phase 1 study has been carried out for the purpose of assessing safety for site inspection visits, sufficient information should now be available to produce a health and safety plan under the CDM Regulations and organise the site visit (see Section 5.4.4).

4) Short-term (immediate) risks may have been identified, in which case it may be necessary to complete urgent remedial works.

5) There is evidence for potential contaminant–pathway–receptor linkages and there is a legal or other obligation to take further action to reduce risk assuming that these are deemed significant.

If items 4 or 5 apply, some form of site investigation is probably required for the purposes of risk assessment. Although the Phase 1 assessments identify whether each site might be potentially contaminated, the actual presence of contamination cannot be determined from a study of historical land-use, walkover visit and pollutant linkage assessments. Site investigations confirm the presence or absence of contamination and its extent and involve physical surveys of the site. The information that has been gathered will be applied to the design of the site investigation. The planning and execution of the invasive site investigation is described in Chapter 5.

4.4 REPORTING

At this stage a report is prepared detail the Phase 1 findings. The content of a typical report is shown in Box 4.4. Box 4.5 indicates some tips for good practice.

Box 4.4 *Content of a typical Phase 1 report*

A typical Phase 1 report might contain some or all of the following:

- records consulted and maps examined
- other documentation made available (such as earlier site investigation reports)
- details of previous sampling points
- description of geology, hydrogeology and related ground information
- identification of statutory involvement (eg enforcement actions such as works notices issued under Section 161A of the Water Resources Act 1991 etc)
- records of interviews with former site owners, users, regulators and other stakeholders – consultation
- identified contaminants, receptors pathways and pollutant linkages
- record of the site reconnaissance
- information on features that may effect or influence future works at the site (eg ecology, landscape)
- interpretation of factual data (where provided)
- the need for any urgent action
- gaps and uncertainties in the data
- supporting information.

Box 4.5 *Tips for good practice*

A search of old maps and plans can reveal a wealth of detail about the activities carried out at in a locality over the years. The assessor should aim to collect the available maps relating to an area. However, beware of gaps in the sequence of mapping that enhance the possibility of missing contaminating activities. The greater the time period for which maps are unavailable the greater the uncertainty.

Do not be too reliant on single information sources.

Visit the site even if the documentary evidence seems conclusive. If subsequent work at the site is planned, a site visit would identify constraints on access to the site.

Some contamination may not be documented. Material dumped at site (such as fly-tipped material) may not be documented, but could present a hazard to be taken into account by the assessment.

During and around wartime, military and other sensitive sites were not shown, or were deliberately misrepresented on maps of the time.

4.5 FURTHER READING

DEPARTMENT OF THE ENVIRONMENT (1994)
Guidance on preliminary site inspection of contaminated land, CLR 2.
Report by Applied Environmental Research Centre Ltd. Volume 1 and Volume 2, HMSO, London.

DEPARTMENT OF THE ENVIRONMENT (1994)
Documentary research on industrial sites, CLR 3.
Report by RPS Group plc. HMSO, London.

DETR Industry Profiles – see Appendix 1.

SIMPSON B, BLOWER T, CRAIG R N, WILKINSON W B (1989)
The engineering implications of rising groundwater levels in the deep aquifer beneath London
CIRIA Special Publication 69 CIRIA, London.

5 Phase 2 risk estimation – site investigations to acquire data

5.1 PURPOSE OF THIS CHAPTER

Site investigation is a major part of the process to characterise the site. Whereas document-based studies, such as those carried out within the hazard identification and assessment stages of risk assessment, determine the *potential* for contamination being present at a site, site investigations provide the opportunity for the *actual* presence of contamination to be determined. The site investigation is normally the first step in Phase 2 of the risk assessment process and is part of **risk estimation** (see Chapter 6).

Physical site investigations normally involve some form of probing, hole excavation, sampling and laboratory analysis. Alternatively, non-invasive techniques such as ground penetrating radar can be applied. CIRIA Special Publication 103, *Remedial treatment for contaminated land – site investigation and assessment* and Project Report 35, *Rapid characterisation of contaminated sites using electrical imaging* give more information on this subject. (Harris *et al* 1995, Onions *et al*, 1996). Sometimes combinations of both are used. The choice of technique is site-specific and no two site investigations are alike.

This chapter describes ways in which the acquisition of physical data enhances the risk assessment and increases confidence in the outcome of the assessment process. It discusses the importance of collecting sufficient, relevant and reliable data and looks at key considerations in collecting this data. It is not intended that this chapter should provide a detailed description of how detailed investigation strategies are prepared or the techniques of site sampling and testing. Rather, the reader is directed towards the further reading section at the end of this chapter for advice on these matters.

5.2 HOW SITE INVESTIGATIONS ENHANCE RISK ASSESSMENT

In all risk assessments there is an element of uncertainty about the conclusions that are reached (Section 4.3.5). Some of this uncertainty derives from information on contamination contaminants, pathways and receptors. By only considering the *potential* contaminants, receptors and the links between them, the previous phase of investigation may in some way have failed to identify the presence of a particular contaminant, perhaps because its likely presence was not documented, or perhaps a potential pathway was not suspected. Consider the example in Box 5.1, which describes a situation where a contamination problem arose from undocumented contaminant sources and pathways.

If all the relevant contaminants, pathways and receptors are not identified, levels of uncertainty in the risk assessment outcome become high and, as in the case summarised in Box 5.1, the assessment findings become unreliable. In the Box 5.1 example, had the existence of the underground tanks been properly determined, site investigations could have been directed towards the interface of the two sites and the contamination of the groundwater might have been measured. This would have considerably reduced the uncertainty. As was discussed in the previous chapter, physical site investigations may not be required for every situation, because the objectives of the assessment have been met in Phase 1. However, the example illustrates that there is a certain minimum level of information, below which the assessment becomes unreliable. The provision of physical data from site investigation helps to ensure that significant omissions are avoided.

So what is this minimum level? The assessor should decide, by considering the questions below, whether more information would increase confidence in the findings:

- would the acquisition of physical data significantly reduce technical uncertainty?
- would improvements in the technical confidence balance the additional costs of site investigation?
- in view of information already gathered, or the assessor's suspicions, could additional information change the assessor's conclusions about the risks?
- what would be the consequences of failing to characterise a pollutant linkage, perhaps in terms of delay to programme, investment decisions, professional reputation of consultant etc?

Box 5.1 *Consequences arising from incomplete identification of pollutant linkages*

For some years, the underground tanks storing solvents at a factory had leaked. The loss of solvent was small and went undetected for at least ten years. The leak was finally discovered when work began on an adjoining site to build a new warehouse on previously undeveloped ground. Hydrocarbon-laden groundwater entered foundations for the new building. Before construction, a Phase 1 risk assessment had been commissioned. This assessment had concluded that, as undeveloped greenfield land, the site had no plausible pollutant linkages, so no physical site investigation, beyond that needed for foundation design, was required.

The risk assessment had failed to identify both the storage and use of solvents at the adjoining factory and the presence of pipelines carrying the solvents across the boundary. These pipelines, laid in shallow trenches filled with pea gravel, ran alongside an electricity cable, also laid in gravel, the route of which then changed course and crossed the undeveloped site. Both of these cable trenches provided a pathway for groundwater migration that carried the leaking solvent into the vicinity of the new building foundations. The pipelines and cable were privately owned and not documented, although the tanks were recorded on local authority registers.

The construction programme was delayed while actions were taken to deal with the source of the contamination and the affected areas.

5.3 CONFIDENCE IN THE DATA

The assessor should be aware of possible sources of uncertainty throughout the risk assessment process, not least those that may arise through sampling and testing. Uncertainty arises from:

- distribution of contamination about a site – heterogeneity
- misunderstandings about the behaviour of contaminants around the site
- inappropriate sampling procedures and analysis
- insufficient frequency of sampling when the contaminant form or concentration varies over time
- reliability of the test results.

Since it is not possible to eliminate all possible uncertainties beyond doubt, assumptions are made to compensate for gaps in the information. For example, distribution of contamination is unlikely to be even on any site, so contaminant concentrations in the soil are likely to vary greatly. The Phase 1 study will have given some indication of likely distribution of contamination, but careful choice of sampling location is essential. It is common to assume that contamination will be found within particular soil strata, or in particular parts of the site. The assessor should also be aware that contamination could be present as concentrated areas or "hot spots". Sampling techniques are discussed in Section 5.4.5.

Uncertainty may arise if samples are collected at an inappropriate horizon. The distribution of contamination is particularly significant with soil sampling, less so with water samples where an individual sample is more likely than soil to be representative of the medium sampled, especially where the contaminant is readily miscible with water.

But what if assumptions about the site characteristics are incorrect? If site sampling was targeted at the upper 6 m of ground, how would the outcome of the assessment be affected if contamination was found in significant concentrations at depths greater than 6 m below the site surface? Good practice in contaminated land risk management is to make decisions based on the available information, while deciding whether the assumptions made where this information is patchy or absent are reasonable in the circumstances; such decisions are rarely completely free of these sorts of uncertainty.

The assessor should consider how representative sampling and testing is of the actual situation at site and how this affects uncertainty. It should be borne in mind that when samples of soil are taken for testing from a "typical" site, the amount tested is a tiny portion of the total mass of soil (Box 5.2).

Box 5.2 *Sample size versus total soil mass*

> Consider the case where samples of soil are being collected at points along a 25 m grid across the site. The Phase 1 studies indicate that the contamination could be located within 6 m of the surface. Within one square of the grid, the total mass of soil will be in the order of 5×10^6 kg. However the sampling exercise at best would sample four locations within the grid amounting to around 10–12 kg of soil. Of this around 1–5 g of soil would be subject to analysis, so in effect only around 0.0000001 per cent of the heterogeneous soil mass is being analysed.

In the case of water and gases, uncertainty may arise from failing to allow for seasonal effects for when sampling is carried out (for example with landfill gas, where generation rates of gas are depressed during winter periods relative to the summer months).

The assessor should consider all the uncertainties that might arise from the data collection process and try to minimise uncertainty so far as possible at all stages of the assessment. Such considerations should put the conclusions about risk into perspective.

5.4 DEVELOPING AN INVESTIGATION STRATEGY

5.4.1 Introduction

The primary purpose of the investigation is to confirm that the contaminants identified by the findings of the previous phase of study are actually present, where in (or on) the site they are present and how much of each is present. To find this out it is necessary to develop a sampling and testing strategy.

Site investigation usually involves considerable expense, and if the investigation is poorly planned considerable resources may be wasted. Before any investigation, sampling or testing takes place, therefore, consideration should be given to the type, quantity and quality of the data required. Some of the published guidance that is available for planning and executing site investigation is given at the end of this chapter.

Considerations at this stage will include:

- whether the investigation to be carried out in "one go" or staged and whether this a preliminary or full investigation
- what information are the interested parties expecting from the investigation? – perhaps enough information to enable a stakeholder to assess their liabilities, as a step towards planning remediation, or to refine the risk assessment giving a higher degree of confidence than that achieved from Phase 1
- could the contamination investigation be integrated with other site investigations (eg geotechnical, ecological).

Now is also the time to think about results:

- what results are required
- limits of detection
- area to be examined
- over what period will results be collected – what ongoing monitoring might be required
- when are the results needed?

5.4.2 Staging site investigations and "zoning"

Site investigations are often carried out in more than one stage especially when the site history has not been well established by the Phase 1 study. The first stage would be an exploratory investigation, whereas the second stage would concentrate on any contaminated areas found during the first stage (for example in delineating any areas with especially high concentrations of contamination that may have been found). This has the advantage of saving the costs of over-investigation of areas where there is no significant contamination.

The site may also be divided into zones for investigative purposes. These zones are areas of the site that are likely to differ in the nature and distribution of contaminants or their physical attributes. In each zone the type of samples taken, the frequency and the testing carried out may vary and be guided by the nature and distribution present.

5.4.3 Consultations

Some thought must be given to who else might be interested in plans to investigate the site. Obviously the owners and occupiers need to be consulted and access arranged. It may be necessary for the assessor to contact both parties to ensure all access arrangements are acceptable. This must be handled sensitively since confidentiality is often a priority for all enquiries. It may also be necessary to talk to the Environment Agency, especially if the works may affect groundwater or will take place in the vicinity of a watercourse. Other bodies should be consulted as appropriate such as utilities companies (to mark out service runs) or British Waterways Board (for work close to canals). In all these consultations it is usual to have at least an outline sample point plan and idea of the site investigation technique to be used.

5.4.4 Health and safety

At the planning stage it is essential to consider the precautions to be taken to ensure the safety of all involved in the site investigation. It is advisable to make a pre-investigation reconnaissance site visit (if this has not already been undertaken as part of the Phase 1 study) to identify obvious hazards such as asbestos fibre or drums on the site surface, as well as looking at barriers to site access. Safety procedures for site workers should be exhaustive where the likely risks are poorly understood. The appropriate legislation that may apply is extensive, incorporating the general provisions of the Health and Safety at Work etc Act 1974. The CDM (Construction Design and Management) Regulations 1994 may also apply. Box 5.3 gives particular site safety considerations.

Box 5.3   *Items for inclusion in the site safety plan*

A site safety plan should include at least the points below.

a) A description of the main contaminants that are likely to be encountered, and precautions to be taken.

b) Hygiene. Good standards of personal hygiene include the obvious – washing of hands and exposed areas of the body, no eating/drinking at site etc. Sometimes this may need to be supplemented with the use of "antitoxic" barrier creams and site-specific practice prohibiting the removal of contaminated clothing from site. Showers and decontamination units (see HSE document HS(g) 66) may occasionally need to be provided. Protective clothing will also be required. As a minimum this will include gloves, boots and hard hats.

c) Physical hazards. Those likely to be encountered include:
   - toxic/explosive flammable gases. It is important to use intrinsically safe tools where the presence of flammable gases is suspected. No work in confined spaces should take place without prior comprehensive gas monitoring
   - work in trenches and trial pits. In trial pits, trenches or excavations deeper than 1.2 m, support should be provided to prevent collapse occurring. Other personnel should be present outside the excavations at all times. Work in poorly ventilated confined spaces should be avoided at all times. Equipment should be available on site to monitor flammable or toxic gases, alternatively staff can wear gas alarms (see CIRIA 1996)
   - fire. No fires should be started at site and smoking should be prohibited, especially at landfill sites, where underground fires can start easily
   - live utilities, including electrical wires/gas pipes. Statutory undertakers' records should be consulted before entering site as part of the Phase 1 study
   - underground tanks and pipes. These may be encountered on old factory or industrial sites and may not be visible from the surface
   - other physical debris such as rusty nails, broken glass that may cause injury
   - unstable ground
   - wartime ordnance.

d) Contact details in case of emergency (hospitals, ordnance disposal, senior contractors' staff).

5.4.5 Site investigation techniques

Types of equipment

The techniques available for site investigation include intrusive and non-intrusive techniques, that is those that disturb the ground and those that do not. Non-intrusive techniques include remote sensing (such as thermal imagery) and other geophysical methods (such as ground-penetrating radar).

The main types of intrusive techniques are:

- trial pits
- boreholes
- window sampling.

Non-intrusive techniques can be useful since these can provide continuous data over a site rather than the spot information provided by a series of exploratory holes. Also they do not disturb the contamination in the ground and so offer a measure of safety to those carrying out the investigation. Non-intrusive techniques are commonly used to identify irregularities or hidden features below ground such as:

- the presence of disturbed ground
- buried foundations, drums, services or munitions
- changes in groundwater conditions.

For details of the full range of available exploratory techniques, refer to BS 5930 (1999) and the sampling techniques section of BSI/ISO Draft BS 7755, Part 2 (1995).

Choosing contractor, setting specifications and specifying equipment

All ground investigation work should only be carried out by organisations having the necessary specialist expertise, personnel and equipment. The assessor should be careful to define the exact requirements within an agreed specification. Some items to be considered for inclusion in the specification are given in Box 5.4.

Box 5.4 *Items to be considered when choosing site investigation techniques*

When specifying plant suitable for the investigation the following should be considered:

- will the investigation involve excavation of trial pits or construction of boreholes, or both?

If trial pits are to be excavated:

- how deep are these to be?
- what ground conditions will be encountered?
- are any obstructions (eg foundations) expected?
- is there suitable access to site for pitting equipment?
- will entry into deep excavations be necessary?

If boreholes are to constructed:

- is accurate logging of the boreholes required?
- what materials are likely to be encountered?
- is instrumentation (eg gas monitoring standpipes) to be installed? Note that type of instrument will depend upon the accuracy of measurement required and the time periods over which measurements will be made
- which drilling technique is to be employed (shell and auger, continuous flight auger, other)

For both types of excavation, is it required that the position of the borehole/trial pit is to be recorded?

These requirements should be communicated to the contractor to enable them to choose equipment appropriate to the aims of the investigation.

For guidance on the selection of SI techniques, refer to the further reading section at the end of this chapter.

5.4.6 Sampling

General

The location and depth at which the samples are taken will depend very much on the objectives of the study and must always be related to the conceptual model, the physical conditions identified at site and the aims of the investigation. Note that it is easy for over-sampling and high analytical costs to occur. A skilled assessor should be able to obtain sufficient samples to give a good indication of the contamination profile, aided by both visual and olfactory observations. The skill of the assessor is important in the selection of samples and the recognition of undisturbed strata. It is usual to obtain more samples in the field than are usually tested. Remember that it is generally expensive to return to take more samples.

There are many documents which will provide further information, especially BSI (2000, in draft), Ferguson (1993), DETR (1994). Other documents will be found in Appendix 1.

Soils

The British Standard Code of Practice (BSI, 2000, in draft) outlines the two principal approaches to soil sampling. These are:

- targeted (judgmental) sampling which involves focusing the sampling on known, or suspected, point source or areas of contamination
- non-targeted investigation aimed at characterising the contamination status of a defined area or volume of a site or zone, using a regular pattern of sample locations.

Targeted sampling involves taking samples at locations that have been chosen on the basis of the conceptual model. The earlier stages of risk assessment will have identified the likely locations of contamination and possible migration pathways and the sampling locations have been chosen to confirm this. Targeted sampling is useful when point sources of contamination are suspected, such as buried waste deposits or tanks, and may be useful when the boundary of a known area of contamination needs to be defined. Targeted sampling is usually employed when investigation is carried out in more than one phase, particularly at the initial phase where it is used to demonstrate the presence of contamination to confirm the conceptual model.

Non-targeted investigation is usually employed when the location and distribution of contamination is unknown. The sampling locations are distributed according to a defined pattern. This pattern might consist of a regular square grid, a herringbone pattern (where the points on a square grid are offset) or a stratified random pattern (where the site is overlaid with a grid and the location of the sample within that grid is chosen at random). Harris *et al* (1995) describes these in some detail. This approach is adopted when it is considered that there may be "hot spots" of contamination at an unknown location within the site. Research studies has suggests that an approach that identifies "hot spots" will be effective in characterising the whole site. In practice a combination of non-targeted and judgmental sampling is used.

The DETR (1994) provides guidance on sample point location.

Water sampling

Developing a correct strategy for water sampling is as important as that for soils since water plays a major role in the transport of contaminants from their source towards potential receptors. As with soils, water-sampling locations should be guided by the conceptual model. The model will have been compiled by reference to information on the surface and groundwater flow direction and hydraulic properties. Sometimes the objective of water monitoring and sampling will be to confirm these properties, at others it will be to confirm contaminant migration, or both. Often, it is required to determine whether contaminated land is having an adverse affect on water quality. To this end it is usual to sample both on and off site and up and downstream of the contamination to determine comparative water quality (in terms of the presence and concentration of contaminants). Water sampling is also required to track pollution plumes, groundwater contaminant movements, and their effects on potential receptors, to determine "baseline" conditions (conditions prior to disturbance of ground) and the stability ("environmental fate") of the contaminants present. BS6068, Sections 6.4 (1987) and 6.6 (1991) and Harris *et al* (1995) provide guidance on surface water sampling.

The frequency of sampling will depend on the objectives of the investigation. Sampling and in situ monitoring normally occurs on several occasions over a period to assess changes, trends and typical characteristics of the water body and is adjusted to suit the characteristics of the water regime. Where the changes are rapid weekly, daily or even hourly, monitoring may be required. Otherwise monthly or quarterly sampling is usual.

Sometimes there is a risk that inserting monitoring instrumentation into ground can create new pathways for contaminant movement. The possibility that this could occur should be considered. This is particularly important where contaminant-containing water is somehow separated from other water bodies, for example in the case where contaminated groundwater is perched above another groundwater body.

Some water samples deteriorate rapidly after sampling. Therefore on-site or rapid analysis of certain parameters is sometimes necessary (Table 5.1), such as for the measurement of dissolved oxygen in water samples, or for testing volatile species (eg benzene) at site by the use of portable testing kits. Where portable analysis equipment is used, it is essential to maintain the accuracy of the test equipment by regular calibration checks. Generally site test kits cannot be considered as accurate as the corresponding laboratory test. Also, as a general rule, site tests cannot as easily be subject to quality assurance controls, although they provide important and useful information.

Table 5.1 *A general guide to the necessity for testing at the location of sampling*

Requirement	Parameters
Parameters where analysis is usually required at point of sampling	pH, electrical conductivity, redox potential, temperature, dissolved oxygen
Parameters where analysis is only usually required at point of sampling if sample preservation is not possible or delays in sample reaching laboratory	Ammoniacal nitrogen, nitrate, nitrite, sulphide, volatile organic compounds.
Parameters where analysis may be carried out cost effectively using portable equipment for later confirmation by laboratory techniques.	Hydrocarbons by portable gas chromatograph. Soil gases by infrared or electrochemical cell. Soluble metals, chlorides and other anions by colorimetric techniques.

Soil gases

Monitoring of soil gases is often carried out at and near landfill sites, at other contaminated sites where contaminated vapours may be present and at sites where natural soil gas generation (from local geology) is anticipated. Normally analysis includes that for the "bulk" gases methane, carbon dioxide and oxygen, but may also include trace gases or vapours. Portable monitoring equipment is commonly used. Generally soil gas concentrations are time-dependent and monitoring results are affected by many seasonal factors such as temperature and groundwater levels. It is usual to use site measurements of soil gas concentrations and flow rates occasionally subject to confirmation by laboratory testing. Specific guidance for soil gas monitoring strategies will be found in Appendix 3.

Dusts and airborne contaminants

Sampling of dust and vapour is often carried out to determine the volume of such contaminants that might be carried into adjoining property, perhaps creating a nuisance. The nuisance element of contaminated land investigation and remediation must always be considered. In particular, local residents living in the vicinity of a contaminated site frequently express concerns over the amounts of dust, vapours and fibres to which they may become exposed, for example should the site eventually be disturbed for remedial purposes. It is useful to establish background conditions prior to major disturbance of the site, or else simply to monitor airborne contaminants from a site in its current state.

Simple gravemetric measurement techniques are used to measure deposition of dusts at strategic points of the site or beyond the site boundary. More complex techniques are employed to assess airborne concentrations of vapours, which might involve sampling of air by passive absorbance of the vapours followed by laboratory analysis, or else active filtration of air to allow measurement of accumulated airborne fibre.

5.4.7 Laboratory analysis

Laboratory analysis is one of the more costly elements of risk assessment and the one most often requiring the assessor to seek specialist advice. This specialist knowledge is often required to advise in the selection of appropriate analysis techniques, limits of detection and required accuracy.

At the outset it is important to establish good communication and a good working relationship with the chosen laboratory. When selecting a laboratory, the assessor should consider whether the laboratory is nationally accredited, has the ability and capability to process the number of samples that wsill be generated and has a track record in the type of testing proposed. Specifically, the chosen laboratory should:

- have the ability to analyse the samples accurately, with the required precision, within the require timescale
- use an appropriate method (British Standard or other approved or agreed in advance)
- participate within national accreditation schemes such as NAMAS (operated by UKAS) or similar
- operate appropriate quality control and assurance schemes
- operate chain-of-custody schemes to enable samples to be located at all stages of handling and analysis.

The assessor should agree with the laboratory the testing procedures that are proposed and should decide whether these are appropriate to the samples being tested. The units in which the results are to be reported and the limits of detection required should be agreed. In return, the laboratory needs to be made aware of any particularly hazardous or potentially hazardous samples that need particular care in handling – those that might be explosive or flammable, release contaminated dusts or vapours or might otherwise present a hazard to laboratory staff. FOCIL (1999) has prepared standard forms for requesting information on laboratory services and for the laboratory's response.

5.5 CONCLUSIONS

Site investigation plays an important role in proving the actual existence of contaminants, pathways and receptors. However site investigations tend to be subject to budget, time and other constraints, so site investigation data always have an element of uncertainty attached affecting confidence of risk decisions made with data. Generally, the assessor is required to have high levels of technical understanding and competence if valid site investigations are to be planned and executed effectively. However, legal, practical and health and safety issues will arise, and these issues will all have to be addressed during the in the course of typical site investigations. Careful planning will ensure resources are not wasted and the quality of the data obtained is enhanced.

Box 5.5 gives some tips for good practice.

Box 5.5   *Tips for good practice*

> When developing a strategy for groundwater sampling bear in mind that dense non-aqueous-phase liquids (DNAPLs) may also move in a different direction to that of groundwater since DNAPL migration is also dependent on geological factors.
>
> Some physical site testing may be appropriate. For example, it is useful to determine the hydraulic permeability of geological strata underlying the site. This information can confirm the conceptual model by demonstrating the relative hydraulic conductivity of the stratum, so that estimates of groundwater flow rates and contaminant transport can be made.
>
> Sampling tools should be decontaminated between each sample to avoid cross contamination.
>
> During site investigations, care should be taken in the choice of sample storage container (which should be unreactive to the substance being stored) and particular attention given to the use and choice of preservation technique (especially important for waters).
>
> A skilled assessor will be able to draw conclusions from observations made during the site investigation. It is good practice for the assessor to visit the site as much as possible during the investigation (a full-time presence would be ideal). If the data gathered remains unchecked, errors may develop in the interpretation. Consider the soil gas-monitoring standpipe in Figure 5.1. The data reported by the individual appointed to obtain readings of soil gas concentrations within this monitoring standpipe all indicated that gas concentrations were "near atmospheric". A visit to the site by the assessor found that the response zone of the standpipe was partially above ground, so that the soil gas readings would have been considerable diluted by ambient air. For comparison Figure 5.2 shows a diagram of a typical gas-monitoring standpipe.
>
> It is common practice to obtain specialist advice and help with choosing and interpreting the results of laboratory testing.

Figure 5.1 *Soil gas monitoring standpipe apparently indicating ambient air gas concentrations (the slots should not be visible above ground (see Box 5.5)*

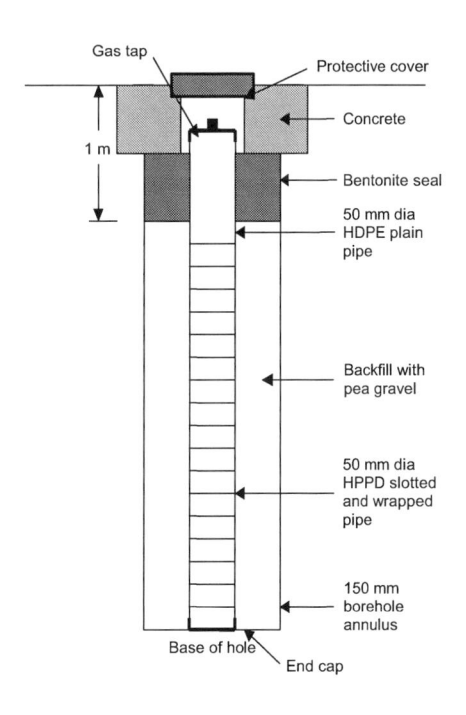

Figure 5.2 *Example of an appropriate soil gas monitoring standpipe (refer to Box 5.5)*

5.6 FURTHER READING

Detailed investigation and sampling strategies

BRITISH DRILLING ASOCIATION (1991)
Guidance Notes for the Safe Drilling of Landfill, Contaminated Land and Adjacent areas
BDA Brentwood, UK

BS 5930 (1999)
Code of Practice for Site Investigation
British Standards Institution, London

BS 6068-6.4 (1987)
Water Quality. Sampling. Guidance on sampling from lakes, natural and man-made
British Standards Institution, London

BS 6068-6.6 (1991)
Water Quality. Sampling. Guidance on sampling from rivers and streams
British Standards Institution, London

BS 6068-6.11 (1993)
Water Quality. Sampling. Guidance on sampling of groundwaters
British Standards Institution, London

BS 6068-6.12 (1996)
Water Quality. Sampling. Guidance on sampling of bottom sediments
British Standards Institution, London

BS 6068-6.14 (1998)
Water Quality. Sampling. Guidance on quality assurance of environmental water sampling and handling
British Standards Institution, London

BS 6068-6.15 (1998)
Water Quality. Sampling. Guidance on preservation and handling of sludge and sediment samples
British Standards Institution, London

BRITISH STANDARDS INSTITUTION (2000)
Revision of BS DD175 Investigation of Potentially Contaminated Sites – Code of Practice, 3rd draft
British Standards Institution, London

CIRIA (1996)
A Guide for Safe Working Practices on Contaminated Sites, Report 132
CIRIA, London

CROWHURST, D and MANCHESTER, S J (1993)
The Measurement of Methane and Other Gases from the Ground, Report 131
CIRIA, London

DETR (1994)
Sampling Strategies for Contaminated Land, CLR Report No 4
HMSO, London

FERGUSON, C and ABBACHI, A (1993)
"Incorporating Expert Judgement into Statistical Designs for Contaminated Sites"
Land Contamination & Reclamation, vol 1, no 3

FORUM ON CONTAMINATION IN LAND (1999)
Standard Forms for Requesting Laboratory Services and for the Laboratory Response

FOCIL, UK
HARRIS, M R, HERBERT, S M, SMITH, M A (1995)
Remedial Treatment for Contaminated Land, Volume 3: Site Investigation and Assessment, Special Publication 103
CIRIA, London

HEALTH AND SAFETY EXECUTIVE (1991)
Protection of Workers and the General Public during Development of Contaminated Land, HS(G)66
HMSO, London

NATIONAL HOUSE-BUILDING COUNCIL (1998)
Chapter 4.1 "Land Quality – Managing Ground Conditions", *Standards*
NHBC, UK

ONIONS, K R, JACKSON, P J, DAWSON, S C J (1996)
Rapid characterisation of contaminated site using electrical imaging
CIRIA Project Report 35

ONIONS, K R, WHITWORTH, K, JACKSON, P (1996)
"Application of geophysical methods to site investigations at contaminated old collieries"
Quarterly Journal of Engineering Geology, 29, pp 219–231
The Geological Society, UK

REYNOLDS, J (1992)
The use of sub-surface imaging techniques in the investigation of contaminated sites
Proceedings of the 2nd International Conference on Construction on Polluted and Marginal Land, pp 121–131, Engineering Technics Press

REYNOLDS, J (1994)
Resolution and differentiation of sub-surface materials using multi-method geophysical surveys
Proceedings of the 3rd International Conference on Reuse of Contaminated Land and Landfills, pp 63–67, Engineering Technics Press

REYNOLDS, J (1998)
The role of geophysics in the investigation of contaminated land
Proceedings of the 5th International Conference on Reuse of Contaminated Land and Landfills, pp 131–137, Engineering Technics Press

SCOTTISH ENTERPRISE (1998)
How to Investigate Contaminated Land, requirements for Contaminated Land Site Investigations, 2nd edition
Scottish Enterprise, Glasgow

WELSH DEVELOPMENT AGENCY (1993)
The WDA Manual on the Remediation of Contaminated Land,
ECOTEC & Environmental Advisory Unit Ltd
Appropriate Health and Safety Precautions

6 Phase 2 – estimation and evaluation: the significance of risk

6.1 INTRODUCTION

Estimation and evaluation is the process of determining whether a site presents a risk and the significance of the risk. There are two stages:

- risk estimation
- risk evaluation.

The activities within each of these two stages are as below.

1. **Risk estimation** is the process of estimating the risks that defined receptors will suffer harm. This involves the consideration of the likelihood, nature and extent of exposure and the magnitude and extent of the effects should exposure occur. This stage includes site investigation (see Chapter 5).

2. **Risk evaluation** is the process of evaluating the need for risk management action, with regard to the magnitude of the risks, the level of uncertainty and, if remedial action is needed, the objectives and the broad costs and benefits.

In turn, two main types of risk are generally considered.

1. **Short-term (acute) risk** is the probability that adverse effects will occur as a result of short term exposure to, or contact with, hazardous substances, or as a result of short-term hazardous conditions.

2. **Chronic (long-term) risk** is the probability that an adverse effect will occur as a result of long-term exposure to, or contact with, a hazardous substance or as a result of a long-term hazardous condition.

The estimation and evaluation process is summarised in Figure 6.1.

The conceptual model is key to all stages of contaminated land risk assessment. The conceptual model produced at Phase 1 stage should be refined with the information from any site investigation which are likely to have demonstrated the existence and nature of contaminants, pathways and receptors. The likely pollutant linkages should be examined and, if found to be still plausible, they should be re-examined in more detail. It is useful to tabulate the likely pollutant linkages at this stage.

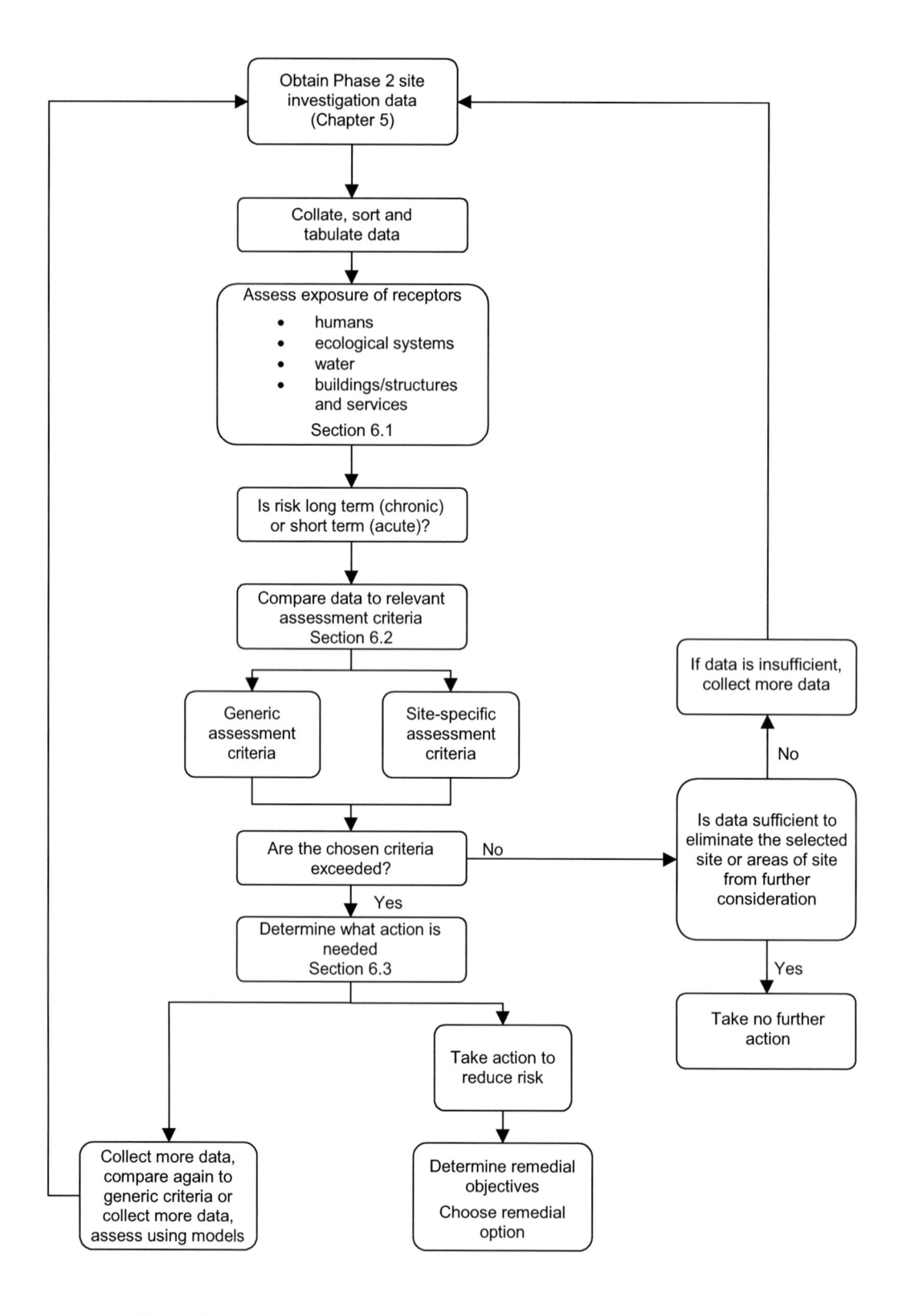

Figure 6.1 *Estimation and evaluation of risk from site investigation data*

It is preferable to use site investigation data (see Chapter 5) since this provides real information about site conditions. In some situations, however, estimations can be made without site investigation data. Box 6.1 describes examples where site investigation data was not required. It should be noted that while an estimation of risk may be possible in some situations, site investigation data is usually required to determine the most suitable, and cost-effective, remediation technique, should remediation be required.

Box 6.1 *Examples of risk estimation without site investigation data*

> **Example 1 – gasworks site**
>
> A Phase 1 desk study was conducted, which revealed that a site had been in operation as a gasworks until the early 1960s. From the site walkover, there was visible contamination present, the characteristic blue staining of "blue Billy" (typical gasworks waste, high in cyanides and sulphate) was observed on bare unvegetated fill, and tar was seen seeping from an embankment. From the geological mapping, it could be seen that the site was underlain by at least 15 m of clay, overlying a minor aquifer. The assessor concluded that there was a good chance of contaminants being present on the surface of the site in concentrations above generic assessment criteria for the protection of human health. As the existing site use was for informal recreation, it was decided that urgent action was required. This took the form of restricting access to the site.
>
> In this example, it was possible to determine the need for remedial action without detailed information on contamination and to carry out short-term measures to control risk. To determine the longer-term actions required if the site was to be returned to beneficial use demanded further data on the extent of contamination present, the nature of ground conditions etc. Remediation was needed for the planned redevelopment of the site as a shopping complex.
>
> **Example 2 – pre-acquisition desk study**
>
> A pre-acquisition study was undertaken on the site, currently used for offices. The site was developed in the late 1960s, consisting of several office blocks and associated hardstanding and landscaping (mainly in planters). A desk study showed the site to have been a former waste tip of a copper works. The underlying geology is a non-aquifer (clay) and the nearest surface water feature is some 500 m from the site. A site visit indicated that infiltration into the ground from the site was likely to be limited as the site surface was mostly covered by hardstanding, which was in good repair.
>
> The purchaser wished to buy the offices for their business, no change in use of the site was planned. It was concluded, that although the underlying fill was likely to contain elevated levels of contaminants, it was unlikely that they would present a risk to a receptor as the pathways were not apparent and no pollutant linkages were identified. No site investigation was required.

6.2 RISK ESTIMATION

Model Procedures (DETR, 1999 in draft) recommends a tiered approach be taken to contaminated land risk assessment. The tiered approach is described as:

- interpretation of site investigation results using generic assessment criteria (Tier 1) and, where available and appropriate

- interpretation of site investigation results using site specific assessment criteria (Tier 2).

Generic assessment criteria are derived and published by an authoritative body. They take into account generic assumptions about the characteristics of contaminants, pathways and receptors and are designed to be protective in a range of defined conditions.

Site-specific assessment criteria are those derived for individual sites when generic assessment criteria are either unsuitable or unavailable.

Further tiers of assessment may be required, each one with increasing level of certainty, and usually additional data requirements. Each increase in tier will generally increase the level of data and time required to estimate the risk. For example, for Tier 2 it is likely that additional data regarding permeability of ground will be required to assess the potential for contaminant migration. This level of data will not always be the needed for

Tier 1, as Tier 1 involves direct comparison of the contaminant source to generic assessment criteria.

The use of the term "site-specific" in this context is slightly misleading as all contaminated land risk assessment is site-specific, in that the derivation, development and testing of the conceptual model is unique to an individual site. The terms "generic|" and "site-specific" refer to the assessment criteria rather than the risk assessment itself. Generic assessment can also be referred to as "qualitative". Site-specific assessment is often referred to as "quantitative".

At first (Tier 1) site investigation data are compared to appropriate generic assessment criteria (where available). As these guidelines tend to be set at "safe" levels (ie incorporating safety factors to account for uncertainty), if the observed concentrations fall below them, then the individual contaminants and/or areas of the site can be considered not to present unacceptable risks. Box 6.2 gives guidance on how to compare site investigation data to assessment criteria.

Box 6.2 *Comparison of site investigation data to assessment criteria*

The first step in comparing site investigation data to assessment criteria is to establish the averaging area. This is usually done when planning the site investigation, as this is key to the spacing and amount of sampling required (see Chapter 5). The averaging area is that over which receptors are exposed to contaminants. In the case of a housing estate, when considering exposure to contaminants in soil, the averaging area will normally be based on the size of gardens. In the case of a sports ground the averaging area might be the area most commonly used for recreation (for example, the area of a football pitch).

When considering site investigation data it is usual to utilise statistical methods for analysing the data, in particular results of contaminant analysis. The cautious approach is to consider the worst case scenario, so that the highest measured concentrations of contaminants are representative of the site as a whole. Of course, this only works if the site investigation has indeed determined the highest concentrations on site. If the assessor is confident that site investigation data are sufficient, and the maximum concentrations fall below suitable generic assessment criteria then it can be concluded that the site presents no risk.

Where the maximum values exceed the assessment criteria it is good practice to try to assess the spatial variation within the averaging area. *Model Procedures* (DETR, 1999 in draft) suggest that the average concentration of soils within the averaging area should provide a reasonable guide to contaminant intake over a long period. Further degrees of caution can be attained by the use of statistical methods such as the upper confidence level of the mean (USEPA, 1989). It must be stressed that, if using complex statistical analysis, the investigator should seek advice from a specialist statistician.

Care needs to be taken if contaminants at the site are approaching concentrations where they may pose a short-term (acute) risk, as opposed to a long-term (chronic) risk. In this scenario action should be taken to deal with the "hot-spot" of contamination regardless of the averaging area.

The process of comparing average concentrations of contaminants against assessment criteria is termed "acceptance testing". There is currently no standard UK methodology or guidance, but research has been undertaken (DETR, 1999, in draft), on statistical comparison of concentrations of contaminants against assessment criteria.

If appropriate generic assessment criteria are not available and/or if observed concentrations exceed the relevant generic assessment criteria then the assessor may:

- collect additional data and reassess the information using generic assessment criteria, or
- evaluate the need for remedial action, or
- go to Tier 2, and allow site-specific assessment criteria to be generated.

Further tiers of assessment may be required, each involving more detailed examination of the site. The driving force for increasing the level of detail and sophistication is to reduce uncertainty. Reduced uncertainty may lead to lower remediation costs, and/or increased confidence in the remediation scheme. Tier 2 assessments usually require additional data to that for Tier 1 and therefore may need further site investigation. The decision to undertake a Tier 2 assessment is a both a financial and technical decision. It may be more economic to undertake remediation rather than go to the expense of a Tier 2 assessment. Guidance adopting a tiered approach includes Risk-Based Corrective Action (ASTM, 1995 and ASTM 1998) and the Methodology for the Derivation of Remedial Targets for Soil and Groundwater to Protect Resources (EA, 1999).

An estimate of risk can be obtained by comparison of site investigation data to assessment criteria. Where relevant generic assessment criteria exist these can be used. Where these are not available, site-specific assessment criteria can be generated. The following sections detail methodologies for assessing the impact of contaminants upon:

- humans – "commonly encountered" contaminants
- humans – landfill gases and other bulk gases
- humans – asbestos
- humans – biological
- humans – explosives and munitions
- humans – radioactive materials
- water environment – all contaminants
- flora and fauna – all contaminants
- buildings, structures and services – all contaminants.

In addition, the commonly used risk assessment models are examined and summarised.

6.2.1 Humans – commonly encountered contaminants

Generic assessment criteria

For human health, Tier 1 assessment involves comparing concentrations of a contaminant to suitable generic assessment criteria. Care must be taken as many published "guidance values" are either:

- not risk-based
- based on differing countries legislation and agendas
- based upon risks to a specific receptor (eg plants rather than humans), and/or
- out of date.

The following sections describe the generic assessment criteria available for assessing risk to differing receptors and also guideline values, which are often inappropriately applied (mainly the Kelly/GLC values (Kelly, 1980), Dutch Intervention Levels (MHSPE, 1994) and the Dutch ABC Values (Asink and Brink, 1985)).

It is also worth considering, at this stage, whether a risk is considered short-term (acute) or long-term (chronic) (Environment Agency, in preparation). Most generic assessment criteria are based upon chronic risks. When considering short-term risks it is necessary to compare contaminant concentrations directly with toxicity or ecotoxicity data rather than generic assessment criteria of this type.

When the concentrations of a contaminant fall below the appropriate generic assessment criteria it can be surmised that the individual contaminant or area of the site is not significantly contaminated and that it does not pose an unacceptable risk.

However, when the assessment criteria are exceeded it is assumed that the contaminant or area of the site poses an unacceptable risk and that further action should be taken. The action may be to confirm that risk, by, for example, undertaking further investigation to define the extent of the unacceptable concentrations of contamination. Alternatively, it can be to consider some form of remedial action to reduce the concentrations of the contaminant, or break the pollutant linkage by some other means. The criteria are developed so that they are set at "acceptable" levels, usually (but not exclusively) taking into account safety factors and assumptions about exposure.

The rationale behind the choice of generic assessment criteria and limitations to their applicability should be clearly understood (and explained in the report) by the assessor.

In the UK the generic assessment criteria produced are not mandatory. Some other countries with requirements for contaminated land assessment have their own guidelines and standards. Although these have no statutory basis in the UK they are nevertheless commonly used in certain situations in the UK where circumstances permit. As part of this research project a consultation exercise was undertaken, which indicated that the main criteria used in the UK are:

- guidance produced in the UK by the Interdepartmental Committee on the Redevelopment of Contaminated Land (ICRCL, 1983)
- Dutch-derived target and intervention levels (MHSPE, 1994), although sometimes these are used inappropriately.

The consultation for this research established that other criteria are also used occasionally, such as:

- USEPA screening values (USEPA 1996)
- Kelly's or GLC values (Kelly, 1980) – usually inappropriate
- Dutch ABC levels (Asink and Brink 1985) – superseded by Dutch target and intervention levels (MHSPE, 1994).

Of these values, ICRCL and Kelly's are not expressly based upon risk. Kelly's were developed to aid disposal rather than assess risk and are therefore difficult to use as generic risk-based assessment criteria. It is also notable that the Dutch target and intervention levels are based on social and physical assumptions, which are not always relevant to the UK (see Box 6.3).

The use of non-UK derived criteria in the UK situation must be approached with caution. If generic assessment criteria from other countries are used the assessor must understand the assumptions underlying the derivation of the values. For example, the Dutch values require a specific sampling method and calculation to account for factors, such as soil matrix and organic matter. The Dutch values also do not account for background dose.

The assessor should be aware as to how, and in which circumstances, the guidance was intended to be used. Some guidance (eg ICRCL) has been developed to allow for different end uses for a site. Others (eg Dutch) have been developed in the context of a multifunctionality approach (in contrast to the suitable-for-use approach) to contaminated land; that is, for land to be remediated for any end use. The Dutch guidance is derived from a mixture of ecotoxicological and human health toxicological considerations as illustrated in Box 6.3.

Box 6.3 *Illustration of ecotoxicological and human health consideration within Dutch Intervention values for soils*

Van den Berg *et al* (1993) reported the basis of the derivation of the intervention values for soils prepared for contaminated land assessment in the Netherlands. They are based upon two key considerations, ecotoxicity and human health, and maximum tolerable risk values are determined for each of these. Usually the lower of these two maximum tolerable risk values is presented as the intervention value. For example:

Barium
Ecotoxicological intervention value	625 mg/kg
Human health toxicological intervention value	4300 mg/kg
Dutch intervention value	625 mg/kg

Benzene
Ecotoxicological intervention value	25 mg/kg
Human health toxicological intervention value	1.1 mg/kg
Dutch intervention value	1 mg/kg

However, this is not always the case, for example:

Lead
Ecotoxicological intervention value	290 mg/kg
Human health toxicological intervention value	300 mg/kg
Dutch intervention value	530 mg/kg

In the case of lead, the Dutch intervention value was set at a concentration closer to the original Dutch C level (from the preceding Dutch guidance) for political reasons, rather than based upon the ecotoxicological or human health toxicological values.

Therefore, if considering human health the use of the Dutch intervention values for barium may be over-conservative. The use of the Dutch intervention value for lead may be inappropriate when considering human health, as this intervention value is the result of a Dutch political consideration, rather than expressly based upon risk.

In the UK, the ICRCL trigger and action values are to be replaced by new guideline values derived from the Contaminated Land Exposure Assessment (CLEA) model as described in CLR 9 (DETR, in preparation) and CLR10 (DETR, in preparation). When the CLEA values are published they will be, in relation to human health, the regulator's preferred generic assessment criteria in the UK.

The CLEA model has been developed to derive human health guideline values. The CLEA model considers various exposure pathways by which humans can be exposed to soil contaminants. Ten exposure pathways are considered (Table 6.1).

It is important that those using generic assessment criteria produced by CLEA understand the basis of the model, that is the particular exposure scenarios considered (this equally applies to all generic assessment criteria and risk assessment models). CLEA will no always be suitable in all circumstances, for example were the conceptual model does not match with the exposure scenarios within CLEA.

Table 6.1 *The exposure pathways considered in CLEA*

Exposure scenarios

Outdoor ingestion of soil
Indoor ingestion of soil
Consumption of home-grown vegetable
Ingestion of soil attached to vegetables
Skin contact with outdoor soil
Skin contact with indoor dust
Outdoor inhalation of fugitive dust
Indoor inhalation of fugitive dust
Outdoor inhalation of soil vapour
Indoor inhalation of soil vapour

Considered in calculation of background dose	– Ingestion of drinking water from mains supply – Skin contact with mains water during showering and bathing – Inhalation of vapours during showering and bathing and from ambient vapours otherwise derived from the mains

Sometimes criteria are used that have not been specifically produced for contaminated land risk assessment. These include standards for drinking water quality, background concentrations of a particular substance in soil and HSE occupational exposure limits (HSE, 2000). The criteria chosen should be relevant to the contaminant–pathway–receptor scenarios identified in the conceptual model. In particular, the acceptability of these guideline values may depend on whether exposure is voluntary or involuntary, people generally accept a greater risk if exposure to that risk is voluntary, see Chapter 7.

When assessing more than one contaminant it is important to consider the effects of the mixture of contaminants. *Model Procedures* (DETR 1999, in draft) also gives guidance on assessing mixtures of contaminants (Box 6.4).

Site-specific assessment criteria

In some circumstances, when appropriate generic assessment criteria is not available, or not relevant to the contaminant, pathway or receptor, it will be necessary to derive site-specific criteria. There is no site-specific risk assessment methodology that is mandatory in the UK and current practice involves the adaptation of methodologies produced by other countries. Consequently, current practice often seems to vary depending on the practitioner's knowledge and previous experience. The calculation of site specific criteria is complex and it is advised that this task be assigned to a specialist in the field.

Whatever risk methodology is used, they all have the same fundamental basis. Each attempts to determine whether contaminants measured on, or in, a site present a risk to a receptor. This is done by estimating the dose of contaminant(s) to a given receptor. This could, in the case of groundwater be by direct measurement if contamination is thought to be entering the aquifer, or by estimating the likely dose. In the case of human health assessment the latter is the usual method. Guidance on the derivation of these values can be found in the *Handbook of Model Procedures for the Management of Contaminated Land*, CLR 11(DETR, 1999, in draft), the CLEA model (DETR, in preparation) and *Collation of Toxicological Data and Intake Values for Humans* (DETR, in preparation).

Box 6.4 *Risk estimation of mixtures of substances*

> *Model Procedures* (DETR, 1999, in draft) indicate that the effects of interaction between different substances are not well understood. General guidance can be found in *Contaminants in soils: Collation of toxicological data and intake values for humans*, CLR9 (DETR, in preparation).
>
> The effects of mixtures of contaminants are reported to be either:
>
> - independent
> - additive (ie overall effect equal to the sum of the individual effects)
> - synergistic (ie overall effect greater than the sum of the individual effects)
> - antagonistic (ie overall effect less than the sum of the individual effects).
>
> *Model Procedures* also notes that remedial action to deal with one substance may treat other substances which may be having an additive effect. For example, where concentrations of a substance in a soil require the soil to be removed to a landfill, other contaminants within that soil will be similarly treated.
>
> Where the above does not apply the recommended approach is:
>
> 1. Treat carcinogenic substances as additive and reduce the generic assessment criteria by a factor of the number of substances being considered (ie if three carcinogenic substance are found on site, reduce the generic assessment criteria of each by a factor of three). (See Box 6.7).
>
> 2. For other substances:
>
>   a) identify all substances that exceed their generic or site-specific assessment criteria and determine possible remedial requirements
>
>   b) identify any substance not dealt with by a) and which also exceeds 0.5 of its assessment criteria
>
>   c) treat substances as additive (reduce assessment criteria by a factor of the number of substances being considered), unless professional toxicological advice indicates the contrary
>
>   d) re-evaluate possible remedial measures.
>
> Example:
>
> > Substance X has been shown to exceed its generic assessment criteria. Remediation is required. Substances Y and Z are identified as being above 0.5 of the generic assessment criteria and remedial measures planned for Substance X will not address Substance Y or Z. Therefore, reduce generic assessment criteria for all three substances by a factor of three and re-evaluate the need for remediation.
>
> It also noted in *Model Procedures* (DETR, 1999, in draft) that synergistic effects are unlikely at levels of exposure close to the tolerable daily intake.

The following description mainly applies to chronic human health risk assessment. Box 6.5 introduces the terminology associated with site-specific risk assessment.

Site-specific assessment criteria can be generated by estimating the tolerable daily soil intake (TDSI – see Box 6.6) and then using this to calculate the critical soil concentration value (C_{Scrit}). This value can then be designated as the site specific assessment criteria and used in a similar manner to generic assessment criteria.

It must be remembered that many assumptions are made in order to derive site-specific criteria; that is, assumptions of the amount of contaminant that the individual may take into the body via each pollutant linkage, the relative body mass, the duration of exposure to the contaminant and so on. The assessor should carefully consider these assumptions and decide whether they are appropriate to the actual conditions at site.

Box 6.5   *Terminology (Model Procedures, DETR, 1999, in draft)*

Estimated daily intake (EDI) – the intake, or dose, of a contaminant from a site for a relevant pathway.

Total estimated daily intake (TEDI$_{SS}$) – the intake, or dose, of a contaminant from a site for all relevant pathways.

Tolerable daily intake (TDI) – an estimate of the average daily intake of a contaminant, expressed in terms of μg d^{-1}, that can be ingested over a lifetime without appreciable health risk.

Mean daily intake (MDI) – a measure of the background intake (in μg d^{-1}) of a contaminant from ambient concentrations in food, water, and air for the UK population.

Tolerable daily soil intake (TDSI) – the maximum intake of a substance that can be allocated to a contaminated soil (see Box 6.6).

Soil allocation factor (SAF) – the proportion of TDSI that can be allocated to a site.

Soil concentration factor (C$_S$) – the unknown soil concentration value.

Critical soil concentration value (C$_{Scrit}$) – the unknown soil concentration value, usually designated as the site-specific assessment criteria. This value is calculated during the site-specific assessment criteria generation process.

Box 6.6   *Tolerable daily soil intake (from Model Procedures, DETR, 1999, in draft)*

The *tolerable daily soil intake* is a term used to describe the maximum tolerable intake from a contamination source (usually a soil, but can be applied to water) allocated to a site. This is derived from two main values (see Box 6.5):

- the tolerable daily intake (TDI)
- mean daily intake (MDI)

Simply, the TDSI = TDI – MDI

However, there are several other factors to consider:

- *soil allocation factor*. A decision has to be made whether the contaminant source represents the only source of the contaminant to the receptor (other than those already accounted for in the MDI). This is usually a balance of the costs of reducing the risk from any one source against the benefits of doing so
- *zero balance*. In situations where the MDI exceeds the TDI (a good example being benzene), the TDSI is effectively either zero or negative. This effectively means that no dose of contaminant from the site is acceptable. In these cases, the TDSI is usually derived as an increment of MDI, so that the increment is small when compared to the MDI, and even smaller when compared to the actual range of intakes, and would not be detectable in general background variations in intake. In such cases, the TDSI needs to reduce the dose as far as reasonably practical
- *other health-based criteria*. The TDSI may also be derived from other health-based criteria, where extensive data exists (for example, lead)
- *UK approach*. Further guidance on the derivation of TDSI, TDI and MDI are found in CLR9 (DETR, in preparation).

The method of assessing risks of carcinogenic substances in the UK is explained in Box 6.7. Note that the UK approach differs to that adopted by the USA.

When generating site-specific assessment criteria, bioavailability (see Box 6.8) may be a consideration.

Box 6.7 *The assessment of carcinogenic substances*

The assessment of carcinogenic compounds can be problematic. Some will not have a "no-effect level" (NOEL and NOAEL), that is, there is no concentration below which the dose is considered safe.

Carcinogenic compounds can be considered in two categories:

- genotoxic (mutagens)
- non-genotoxic (non-mutagens).

Mutagens are generally considered as not having a no-effect level (CARACAS, 1999, DETR, in preparation).

In the USA, these both genotoxic and non-genotoxic substances are assessed using a slope factor (essentially a measure of the chemical potency in causing cancer) to gauge the potential health risk, although it has been proposed to evaluate carcinogenic substances on a case-by-case basis (USEPA, 1996). WHO favours the use of slope factors for the evaluation of genotoxic substances and uncertainty factors for non-genotoxic substances. Both WHO and USEPA endorse the use of quantitative risk assessment (a specific methodology for the extrapolation of dose/effect relationships).

The UK Committee on Carcinogenicity of Chemicals in Food, Consumer Products and the Environment (COC) does not endorse the use of quantitative risk assessment. However, the UK approach does involve the use, in part, of quantitative risk assessment to generate a TDI for genotoxic compounds. In UK guidance, therefore, a TDI can exist for both genotoxic, non-genotoxic and non-carcinogenic contaminants.

There are also differences in approach to the theoretical tolerable excess cancer risk that is acceptable. An excess lifetime cancer risk is the additional risk, over background exposure, an individual faces of getting cancer from the contaminated source.

Choice of excess cancer risk is a mixture of political and scientific judgement. In the USA the tendency is to adopt a risk of 10^{-6} excess lifetime cancer risk, or lower, as acceptable. Holland uses 10^{-4} for excess cancer risks per substance (note: the Dutch do not consider MDI, ie TDSI=TDI), which equates to a 10^{-6} lifetime risk. In the UK, risks around 10^{-5}–10^{-4} excess lifetime cancer risk are acceptable (HSE, 1999) (CARACAS, 1999). In the UK, carcinogenic substances are treated as additive and the generic assessment criteria reduced by a factor of the number of substance being considered (ie if three carcinogenic substance are found on site, the generic assessment criteria of each is reduced by a factor of three). Specific guidance on how these risks are to be interpreted are generally not available (CARACAS, 1999).

CLR9 (DETR, in preparation) states that, "the most appropriate figure to adopt as an upper limit for an 'acceptable' additional lifetime risk of cancer from exposure to environmental contamination from a contaminated site may be 10^{-4}, that is an annual excess lifetime risk of cancer of about 10^{-6}". It should be noted that while based upon science, the decision determining the acceptability of risk is essentially a consideration of the calculated risk (incorporating uncertainty), practicality, cost and priorities.

Box 6.8 *Bioavailabilty*

> When applying generic assessment criteria, it is generally assumed in the UK that all the concentration of contaminant measured at the source is available to a receptor, once dilution and attenuation are taken into account (CARACAS, 1999). This need not be the case, especially for soil-bound contaminants. Assessors may need to consider:
>
> - *mineral and chemical speciation*, ie the contaminant may be a component of another compound with different chemical properties. Soil properties may have an influence on the mobility of a contaminant
> - *sorption effects*. Certain soils may act as a sponge for contaminants, binding them into the soil matrix (clays in particular)
> - *transit time*. The matrix a contaminant is in has an effect on the residency time in a receptor – eg soil passes through the body slower than water. This has an effect on the uptake of the contaminant within the body.
>
> Some elements of bioavailability are contained in the CLEA model (for example, the soil matrix). Assessors should consider these effects, especially when deriving site-specific assessment criteria.

6.2.2 Humans – landfill gases and other bulk gases

The hazards to human health from gases in the ground are associated with the potential for explosion, poisoning and asphyxiation. The majority of generic assessment criteria apply to the point of exposure rather than the concentration of these soil gases, although for a conservative approach their application to concentrations in the ground can be used to give a rough estimate of risk. Appendix 3, Section 3 details guidance available for risk estimation of soil gases, which requires specialist advice.

The hazards related to explosion and asphyxiation are dealt with in the section detailing risks to buildings and structures.

The effects of poisoning should be treated in a similar way to conventional contaminants. The Health and Safety Executive Occupational Exposure Limits (EH40) (HSE 2000) tend to be used as generic assessment criteria, although when being used to assess risks to the general public it is usual to divide the exposure limit by a factor of 40. Effectively this is to take account of the additional exposure time and increased sensitivity of sub-groups of the general population (when compared to the working population) (CLR9, DETR, in preparation).

6.2.3 Humans – asbestos

No screening threshold can be set for some substances, since they are likely to present risks even at the lowest concentrations. For example, no safe concentration can be proposed for asbestos, because a safe threshold is not known to exist (*Air Quality Guidelines for Europe*, WHO, 1997). Asbestos only has the potential to cause harm if the pollutant linkage is complete. Since asbestos is a hazard to human health only if it enters the lungs, circumstances where asbestos fibre may become airborne are the most important pathway. Therefore, a site should be considered to present a risk if:

a) free fibres of asbestos are measured on site or could arise from other present forms of asbestos in the future; and

b) the pollutant linkage is complete (ie asbestos must be able to reach a receptor).

Further guidance on the topic can be found in Appendix 3, Section 6.

6.2.4 Humans – biological hazards

Hazards to human health arise from micro-organisms present in the soil. Most micro-organisms are harmless to man, although certain bacteria, viruses and fungi are harmful to humans depending on the environmental conditions and exposure. Frequently the risks posed are short-term (acute) and are thus of particular significance during site investigation and construction.

There are no generic assessment criteria available for biological hazards and this type of risk estimation would fall under a Tier 2 assessment. Further information is contained in Appendix 3.

6.2.5 Humans – explosives and munitions

There are no officially recognised generic assessment criteria for explosives and contamination. Hazards are generally the short-term (acute) risk to humans from explosion and fire, and the longer-term (chronic) risk from toxicity.

For chronic toxic hazards, explosives should be treated as other "standard" contaminants. Some assessment criteria do exist although in most cases a Tier 2 assessment will usually be required. The Environment Agency is undertaking research into risk assessment of explosives and munitions (Environment Agency, in preparation). With short-term risk, any potential explosive or potential fire risk is obviously a hazard. Refer to Appendix 3.

It is worth noting that specialist consultancies that deal with this type of contamination routinely should be considered if explosives and munitions are likely to exist on a site. Several of these consultancies have developed in-house generic criteria for the assessment of explosives and munitions.

6.2.6 Humans – radioactive materials

The risk from radioactive materials is both radiological and toxicological. At the concentrations usually encountered outside of specialist nuclear facilities the main risks are likely to be long-term (chronic), with increased cancer incidence being the main radiological risk.

Risk assessment of radiological contamination is a specialist activity and it is advised that specialist advice is sought. For nuclear industry-related scenarios the risk assessment will generally be performed on, or on behalf of, the nuclear industry operator. Outside of the nuclear industry, radioactive contamination is most likely to be encountered as a result of manufacturing processes that have used fluorescent paints and specialist alloys, and as a result of disposal of radioactive sources from x-ray sets or laboratories. Risk assessment is largely similar to that for other types of contamination although, in particular, the dermal and inhalation pathways will usually assume greater significance.

Appendix 3 gives further detail of guidance for radioactive risk assessment and reviews available guidance.

6.2.7 Water environment – all contaminants

The assessment of risks to the water environment should follow the tiered methodology described in "Methodology for the Derivation of Remedial Targets for Soil and

Groundwater to Protect Water Resources" (Environment Agency, 1999), which describes the tiered approach.

Tier 1 – comparison of pore water quality in contaminated soil to relevant criteria such as environmental quality standards, drinking water standards, background water quality etc.

Tier 2 – considers dilution by the receiving water and compares the concentration in the pore water multiplied by the dilution factor to relevant criteria such as environmental quality standards, drinking water standards, background water quality etc.

Tiers 3 and 4 – consider natural attenuation, using simple analytical models (Tier 3) or more sophisticated numerical models (Tier 4).

This is broadly the same as the tiered approach described in Section 6.2 in that each tier represents an additional level of complexity and certainty. The decision about which tier to use rests in part upon financial and technical considerations.

It should be noted that generic assessment criteria for water resource protection are not necessarily the same as those for human health protection. It is quite common for the assessment criteria to be the quality criteria for the receptor water body, for example environmental quality standards. Box 6.9 outlines the main assessment criteria for water resource protection.

Box 6.9 *Assessment criteria – water resource protection*

UK Private Water Supply Regulations 1991
UK Water Supply (Water Quality) Regulations, 1989
UK Surface Waters (River Ecosystem) (Classification) Regulations 1994
UK Quality standards for water to be used for direct abstraction to potable supply
UK Environmental quality standards for the protection of aquatic life
UK Quality standards for fresh and saline waters used for bathing and contact water sports
UK Quality standards for freshwaters required to support fish
UK River Water Quality Objectives
EC Drinking Water Standards
EC Water Quality Standards

6.2.8 Flora and fauna – all contaminants

Estimating the effects of contamination on flora and fauna is more problematic as there is great variability between species in sensitivity to contaminants. Some generic assessment criteria have been developed, such as those relating to phytotoxic risk in the ICRLC guidance and the ecotoxicological derived criteria in the Dutch intervention values. These can act as a useful indication, particularly of phytotoxic effects, but much uncertainty will still be associated with using these criteria in this manner.

In most cases the effects of contamination on flora and fauna will be considered:

- when planning plantings and landscaping as part of a redevelopment
- when an ecological or property (in terms of crops, livestock and domestic animals) receptor is identified as part of Part IIA determination (see Appendix 2)
- when identified as a constraint in an environmental assessment.

When planning plantings and landscaping specialist advice with regard to the types of plants selected should be sought if phytotoxic problems are likely to be suspected or encountered. Risk management options, rather than detailed assessment should be considered at this stage. When planning the development, it may be acceptable (in financial terms to the developer) to allow for a certain percentage of plant losses, rather than investigate and remediate to address just this issue. Other options, like importing clean soil for landscaping may be appropriate.

When an ecological or property receptor is identified during the course of a Part IIA implementation, or as a result of an environmental assessment, it may be more appropriate to undertake ecological surveys to assess whether the contamination is causing harm to the receptor. This could be seen in terms of, population densities, health of the species, absence of species etc. For Part IIA assessments harm to ecological receptors is defined as, "harm which results in an irreversible adverse change, or in some other substantial adverse change, in the functioning of the ecological system within any substantial part of that location" (DETR, 2000). For Part IIA assessments harm to crops is defined as, "a substantial diminution in yield or other substantial loss in their value resulting from death, disease or other physical damage" (DETR, 2000). Clearly, this is a specialist activity and a suitable specialist should be appointed.

Appendix 4 describes in detail the processes involved in ecological risk assessment.

Contaminated sites can lead to the development of rare ecosystems as a function of toxicity and isolation. In some circumstances, the remediation of sites, say for the protection of groundwater, can, adversely affect the ecological balance on site. It is also notable that other environmental stresses can lead to the same symptoms as contamination, and conclusions drawn from examination of symptoms alone should be treated with caution. Mitigation measures may need to be considered for such cases.

6.2.9 Buildings materials and services – all contaminants

Construction materials are susceptible to attack from aggressive substances within contaminated soils, groundwaters and from vapours and gases (see Section 6.2.3 on explosive and asphyxiant risks due to soil gases elsewhere within this chapter). In addition the structure of the building, services may act as pathways for contamination to migrate. Contamination may migrate along trenches within which services are laid, and in doing so may be transferred from the original location of the contaminant to a sensitive receptor. In this case the structure or materials are effectively the *pathway*.

Assessment of risks to construction materials will follow the familiar pattern of identifying the hazard (*contaminant*) that could affect the structure or services, the structure or services at risk (*receptor*) and the linkage between them. In many cases the contaminants will have direct contact with the construction materials concerned, such as contaminated backfill placed around foundations, or where foundations such as piles are placed within aggressive ground. Under Part IIA it is possible to have a contaminant–pathway–receptor linkage without direct contact between the building materials or services such as, migration of landfill gases and the effects of unstable ground conditions (say from volume reductions in fill materials due to chemical and biological processes).

Sometimes the effects of contamination will be obvious. Existing development at site may exhibit signs of damage such as cracking, spalling and discoloration that have been caused by contamination. However, expert assistance will be needed in determining whether the observed distress is caused by ground contamination or other sources. If the site is to be redeveloped using similar construction materials it may be judged that damage is likely to occur to these materials from the identified ground conditions. Where new development is planned on sites where no signs of distress to construction materials are evident, risk may be assessed by comparison of observed soil concentrations to guidance values, where these are available.

In most cases generic assessment criteria are not available and a Tier 2 or assessment above (see Section 6.1) will be necessary to assess the risks to buildings and property. The EA is undertaking research into risk to buildings from contaminated land (Environment Agency, in preparation). More details can be found in Appendix 5.

6.2.10 Risk assessment models

The process of estimating risks by the preparation of site-specific criteria can be assisted by the use of models (such as fate and transport models and exposure assessment models). Exposure assessment models estimate exposure to contaminants and calculate assessment criteria. They also may be used to prepare site-specific remediation goals for soil and groundwater. Some models estimate exposure, others demonstrate the transport of contaminants in media, such as groundwater. Some models trace the fate of the contaminant in the environment, perhaps on a time- or distance-related basis, taking into account contaminant changes by degradation etc.

When using models, the assessor must understand the processes by which the model operates and the uncertainties that are created by the model, and decide whether these are compatible with the objectives of the study and a site conceptual model. For example, a model may make the assumption that contamination in the form of vapours passes vertically through the soil at a defined rate. The assessor must decide whether this is reasonable in relation to the actual situation at a site.

As well as uncertainties associated with the model algorithms the assessor must also account for uncertainties associated with the input data. Another consideration should be the sensitivity of the model (ie do small changes to certain input data drastically change the results).

This section concentrates on risk assessment model packages rather than specific exposure assessment or fate and transport models. The consultation undertaken as part of this research identified risk assessment models in common use within the industry. Most of the commercially available exposure assessment models contain, to a degree, some form of fate and transport modelling function, albeit to a lesser level of complexity (in most cases) than specialist fate and transport models. The majority of these models correspond up to a Tiers 3 and 4 groundwater risk assessment. Uncertainty can be further reduced by the use of specialist contaminant fate models to estimate the amount of substance received by the receptor.

This section does not to validate these models, but aims to summarise the main features and uses of each model and highlight any limitations they may have.

Table 6.2 summarises the principles and methodologies of the selected models.

Further details of the software models, including contact details for distributors are given in Appendix 6.

It is worth noting that the CLEA model (DETR, in preparation) was not available at time of publication and so is not included in this section. The basis of the CLEA model is explained in Section 6.2.2.

Table 6.2 Summary of the principles and methodologies of selected risk assessment model packages

Model	Base methodology	Probabilistic (Y/N)	Base data sets	Sources considered	Pathways considered	Receptors considered	Quality of supporting documentation	Notes
BP RISC	ASTM E1739-95	Yes	USEPA – note use of slope factors for assessing carcinogenic compounds. Also makes use of data from New Zealand	Outdoor air	• Inhalation	Humans – eight pre-defined scenarios plus user-defined scenarios	Comprehensive user manual although on line help limited. The manual includes an independent review of the model	Ability to determine risk-based total petroleum hydrocarbon fractions
				Indoor air	• Inhalation			No scope for sensitivity analysis.
				Groundwater	• Ingestion, dermal contact • Volatilisation during showering • Volatilisation and vapour transport into buildings • Volatilisation and transport to outdoor air			Output of 0.00E+00 whenever toxicological information is not available, not to be confused with no risk.
				Soil	• Ingestion, dermal contact • Volatilisation and vapour transport to outdoor air • Volatilisation and vapour transport to indoor air • Leaching to groundwater and then ingestion/dermal contact • Leaching to groundwater and then volatilisation during showering			RISC model uses a volatilisation model based upon a depleting source, whereas the original ASTM and USEPA approaches used a constant-source model designed for use on petroleum-release sites, but can be adapted for other sites
CONSIM	Based upon EA R&D Publication 20	Yes	UK (eg Water Supply (Water Quality) Regulations, 1989, WHO also included	Soils	• Leaching of contaminants to groundwater	Controlled waters	Good online help	Model specifically designed to assess the controlled water receptor
EA R&D Publication 20[1]	Developed from previous R&D Publications 12 and 13	N/A	N/A	Soils and groundwaters	• Leaching of contaminants to groundwater from soil • Direct contamination of groundwater	Controlled waters	Publication is clear and concise. Accompanying spreadsheet also useful	Specifically designed to assess the controlled water receptor

[1] This publication is not a risk assessment model as such, but it is accompanied by a spreadsheet to aid calculation. It is available from the Environment Agency website: (www.environment-agency.gov.uk/gwcl).

Table 6.2 Summary of the principles and methodologies of selected risk assessment model packages (continued)

Model	Base methodology	Probabilistic (Y/N)	Base data sets	Sources considered	Pathways considered	Receptors considered	Quality of supporting documentation	Notes
RBCA Toolkit – Chemical Releases	ASTM E1739-95 and ASTM PS-104	No	USEPA – note use of slope factors for assessing carcinogenic compounds	Groundwater Soil	• Ingestion • Volatilisation during showering • Volatilisation and vapour transport into buildings • Volatilisation and vapour transport to outdoor air • Contamination of surface waters and exposure via swimming, fish consumption, direct exposure of aquatic species • Ingestion, dermal contact • Volatilisation and vapour transport to outdoor air • Volatilisation and vapour transport to indoor air • Leaching to groundwater and then ingestion/dermal contact • Leaching to groundwater, then volatilisation during showering	Humans – residential and commercial receptor defined Groundwater – possible to set maximum contaminant levels (standards) as receptors. Surface water – possible to set maximum contaminant levels as receptors. Also can use aquatic life protection criteria	On line help and manual, although the user will have to refer to the both ASTM standards for guidance	Ability to determine risk-based total petroleum hydrocarbon fractions Receptors, exposure scenarios and contaminant transport parameters all customisable
RISC-HUMAN	Dutch CSOIL model	No	Mainly uses WHO ADI data	Outdoor air Indoor air Soil Groundwater Surface water Milk, fish, meat and vegetables	• Inhalation • Inhalation • Ingestion, dermal contact, inhalation • Inhalation (showering), ingestion of drinking water • Dermal contact, ingestion (including suspended solids) • Ingestion	Humans, predefined exposure population. User can define population	Extensive online help facility, but CSOIL model is not fully described	Model based upon constant contaminant source
Risk* Assistant	USEPA risk assessment guidance	No	USEPA (IRIS and HEAST) – note use of slope factors for assessing carcinogenic compounds New Jersey and California hazard data	Air Surface water Groundwater Soil Sediment Fruit Vegetables Meat and fish Dairy products	• Drinking water • Showering • Indoor air • Outdoor air • Swimming • Dust and soil outdoors • Dust and soil indoors • Consumption of fruit, meat, fish, dairy products and vegetables	Humans – ten predefined exposure populations based upon US statistics. User has option of defining own exposure population	Fairly comprehensive user manual, although assessor will likely need access to USEPA Risk Assessment Guidance manual. Online help limited	Includes sensitivity analysis Model does not consider dermal contact with contaminated soils

6.3 RISK EVALUATION

The purpose of risk evaluation is to decide whether or not risks are acceptable and to determine the need for remedial action. The acceptability of identified risks may depend on who is considering the risks (see Chapter 7). Ultimately, the decision on acceptability of a risk is a balance of the technical reasoning, practicality, perception and cost-benefit.

This stage involves:

- collation and review of the risk-based information for the site
- addressing uncertainty and its effect on judgements regarding risk estimates
- identification of those risks that are considered unacceptable.

6.3.1 Collating and reviewing risk-based information

At this stage it is useful to summarise all the risk-based information for the site and relate the receptors to the relevant contaminants. In effect, this involves a re-examination of the conceptual model in light of new information. For large sites it may be that the site is subdivided into several zones for clarity and ease of assessment.

6.3.2 Addressing uncertainty

Uncertainty should be considered in terms of:

- whether enough data exists to estimate the risks with an acceptable level of confidence
- identification of assumptions and safety factors used in the assessment.

The assumptions and safety factors incorporated into a risk estimation should be examined, and if uncertainty is considered unacceptable then the risk estimation stage is repeated (ie the collection of more site investigation data, see Section 5.3). The cost and benefit of additional risk estimation needs to be balanced against the need for certainty. For some sites, uncertainty may be acceptable, and the costs of additional risk estimation deemed unnecessary. However, further site investigation data and risk assessment may be necessary to achieve a cost-effective remediation strategy.

6.3.3 Identification of unacceptable risks

The following methodology has been developed from an in-house procedure used by Enviros Aspinwall (not published), submitted during the course of this research. This methodology was in turn developed from the "Guide to Risk Assessment and Risk Management for Environmental Protection" (DoE, 1995) and *Draft Statutory Guidance on Contaminated Land* (DoE, 1996). The method presented is an updated and modified version of the Enviros Aspinwall procedure and represents one possible methodology for presenting and evaluation the results of risk estimation.

This method for risk evaluation is a qualitative method of interpreting the output from the risk estimation stage of the assessment. It involves the classification of the:

- magnitude of the potential **consequence** (severity) of risk occurring (Table 6.3)
- magnitude of the **probability** (likelihood) of the risk occurring (Table 6.4).

Table 6.3   *Classification of consequence*

Classification	Definition	Examples
Severe	Short-term (acute) risk to human health likely to result in "significant harm" as defined by the Environment Protection Act 1990, Part IIA. Short-term risk of pollution (note: Water Resources Act contains no scope for considering significance of pollution) of sensitive water resource. Catastrophic damage to buildings/property. A short-term risk to a particular ecosystem, or organism forming part of such ecosystem (note: the definitions of ecological systems within the Draft Circular on Contaminated Land, DETR, 2000).	High concentrations of cyanide on the surface of an informal recreation area. Major spillage of contaminants from site into controlled water. Explosion, causing building collapse (can also equate to a short-term human health risk if buildings are occupied.
Medium	Chronic damage to Human Health ("significant harm" as defined in DETR, 2000). Pollution of sensitive water resources (note: Water Resources Act contains no scope for considering significance of pollution). A significant change in a particular ecosystem, or organism forming part of such ecosystem. (note: the definitions of ecological systems within Draft Circular on Contaminated Land, DETR , 2000).	Concentrations of a contaminant from site exceed the generic, or site-specific assessment criteria. Leaching of contaminants from a site to a major or minor aquifer. Death of a species within a designated nature reserve.
Mild	Pollution of non-sensitive water resources. Significant damage to crops, buildings, structures and services ("significant harm" as defined in the *Draft Circular on Contaminated Land*, DETR, 2000). Damage to sensitive buildings/structures/services or the environment.	Pollution of non-classified groundwater. Damage to building rendering it unsafe to occupy (eg foundation damage resulting in instability).
Minor	Harm, although not necessarily significant harm, which may result in a financial loss, or expenditure to resolve. Non-permanent health effects to human health (easily prevented by means such as personal protective clothing etc). Easily repairable effects of damage to buildings, structures and services.	The presence of contaminants at such concentrations that protective equipment is required during site works. The loss of plants in a landscaping scheme. Discoloration of concrete.

Table 6.4   *Classification of probability*

Classification	Definition
High likelihood	There is a pollution linkage and an event that either appears very likely in the short term and almost inevitable over the long term, or there is evidence at the receptor of harm or pollution.
Likely	There is a pollution linkage and all the elements are present and in the right place, which means that it is probable that an event will occur. Circumstances are such that an event is not inevitable, but possible in the short term and likely over the long term.
Low likelihood	There is a pollution linkage and circumstances are possible under which an event could occur. However, it is by no means certain that even over a longer period such event would take place, and is less likely in the shorter term.
Unlikely	There is a pollution linkage but circumstances are such that it is improbable that an event would occur even in the very long term

These classifications are then compared to indicate the risk presented by each pollutant linkage. It is important that this classification is only applied where there is a possibility (which can range from high likelihood to unlikely) of a pollutant linkage existing.

This method can be applied with or without site investigation data and can be used to assess the results of either qualitative or quantitative assessment. **It is recommended that the amount of data and basis of classifications are made clear when reporting such an assessment**. It is often possible to undertake this risk evaluation following the Phase 1 stage of the risk assessment. If site investigation and further risk estimation are then undertaken the evaluation can be revised.

Once the consequence and probability have been classified, these can then be compared (see Table 6.5) to produce a risk category, ranging from "very high risk" to "very low risk". The actions corresponding with this classification is given in Table 6.6. A worked example is presented in Box 6.10.

Table 6.3 shows the classification of consequence. To classify the consequence it is important to bear in mind that the classification does not take into account the probability of the consequence being realised (this is considered in Table 6.4). Therefore, for a particular pollutant linkage it may be necessary to classify more than one consequence. For example, the risk from methane build-up in a building presents a risk of harm both to the building and to human health. Both would be classified as *severe*, but the probability, addressed in the next stage of this methodology, may vary (for example, the building may be unoccupied for most of the time, with only occasional visits – eg a pumping station).

The classification of *severe* relates to short-term (acute) risks only. The *medium* classification relates to chronic harm, which can be classed as "significant harm" (if the assessment is carried out for Part IIA purposes. The *mild* classification also relates to significant chronic harm but applies to less-sensitive receptors. The *minor* classification relates to harm which, while not considered "significant", may have a financial implication (eg phytotoxic effects of contaminants on development landscaping).

It is worth noting that, in theory, both a *severe* and *medium* classification can result in death. The differentiation between the two categories is that *severe* relates to a short-term risk whilst *medium* relates to a long-term risk. Therefore the classification of *severe* should indicate that urgent action is required (urgent action may also be required under the *medium* classification, but usually longer-term actions are sufficient).

The classification gives a guide as to the severity and consequence of identified risks when compared with other risk presented on the site. It is not possible to classify an identified risk as presenting "no-risk", rather "very low risk". This is important, as the acceptability of risk may depend on the viewpoint of the stakeholder concerned. It may be necessary to take action to deal with a risk even if classified as "very low", although these actions may not necessarily be required urgently.

Table 6.5 *Comparison of consequence against probability*

		Consequence			
		Severe	Medium	Mild	Minor
Probability	High likelihood	Very high risk	High risk	Moderate risk	Moderate/ low risk
	Likely	High risk	Moderate risk	Moderate/ low risk	Low risk
	Low likelihood	Moderate risk	Moderate/ low risk	Low risk	Very low risk
	Unlikely	Moderate/ low risk	Low risk	Very low risk	Very low risk

Table 6.6 *Description of the classified risks and likely action required*

Very high risk	There is a high probability that severe harm could arise to a designated receptor from an identified hazard, OR, there is evidence that severe harm to a designated receptor is currently happening.
	This risk, if realised, is likely to result in a substantial liability.
	Urgent investigation (if not undertaken already) and remediation are likely to be required.
High risk	Harm is likely to arise to a designated receptor from an identified hazard.
	Realisation of the risk is likely to present a substantial liability.
	Urgent investigation (if not undertaken already) is required and remedial works may be necessary in the short term and are likely over the longer term
Moderate risk	It is possible that harm could arise to a designated receptor from an identified hazard. However, if is either relatively unlikely that any such harm would be severe, or if any harm were to occur it is more likely that the harm would be relatively mild
	Investigation (if not already undertaken) is normally required to clarify the risk and to determine the potential liability. Some remedial works may be required in the longer term
Low risk	It is possible that harm could arise to a designated receptor from an identified hazard, but it is likely that this harm, if realised, would at worst normally be mild.
Very low risk	There is a low possibility that harm could arise to a receptor. In the event of such harm being realised it is not likely to be severe.

Box 6.10 *Example of risk evaluation*

A site is used for car parking. The surface is mainly hardstanding, but the quality is not sufficient to prevent infiltration of rainwater. Site investigation has shown that, underlying the hardstanding, the made ground and groundwater (minor aquifer) beneath the made ground contain raised concentrations of toxic metals. The site investigation also encountered several areas of fly-tipped wastes with very high cyanide content (enough to present short-term risks to human health). One such area, bordered by housing, is used for informal recreation, mainly by children.

Therefore the contaminant-pathway-receptor relationship can be summarised as below.

Contaminant	Pathway	Receptor	Consequence of risk being realised	Probability of risk being realised	Risk classification	Risk management action taken
Fly-tipped material with high cyanide content	Direct contact	Humans, mainly children playing on site	Severe	High likelihood	Very high	Immediate removal of fly-tipped material to suitable landfill facility
Toxic metals, for example arsenic and cadmium	Leaching to groundwater (minor aquifer)	Minor aquifer, no local abstractions	Medium	High likelihood	High	Further groundwater monitoring, including perimeter and removal of hotspots of contamination.
Toxic metals, for example arsenic and cadmium	Direct contact	Site workers and visitors during remediation	Medium	Likely	Moderate	Site health and safety plan made allowance for contamination. Site workers were supplied with personal protective equipment and damping down of the site during dry periods was undertaken during remediation.
Toxic metals, for example arsenic and cadmium	Dust	Site workers Residential properties next door to site Site workers and visitors during remediation	Medium	Likely	Moderate	It was considered that damping down of site was sufficient to break this pollutant linkage. Dust monitoring was undertaken on site and at site boundaries to prove this.

Note

The pollutant linkage for residential properties was not assessed in detail, as the measures to address the risk to site workers from contaminated dust were considered sufficient to protect nearby residents.

6.4 REPORTING

The minimum reporting requirements for a Phase 2 risk assessment report are presented in the *Handbook of Model Procedures*, Vol III, (DETR 1999, in preparation).

In summary it is recommended that the following headings be considered:

1. Introduction

2. Background to the assessment – details of Phase 1 assessment

3. Objectives of the assessment – management context, general approach, risk/receptors considered

4. Site investigation (data collection) details – rationale, methodology, specifications, sample handling, site QA/QC, limitations, rationale for selection of tests, sample preparation and analysis

5. Summary of site investigation and analysis findings, site observations, in-situ testing, data, laboratory QA/QC, identification of invalid data, data summary

6. Data interpretation – approach (generic, site-specific) and validity to situation, exposure scenarios, possible effects, selection/derivation of assessment criteria, limitations/constraints, outcome for each receptor

7. Risk evaluation – collation and review of risks, uncertainty, consequences of courses of action, broad costs/benefits, provisional remediation targets.

8. Conclusions and recommendations – acceptability of risks, recommendations for further action.

9. References

6.5 FURTHER READING

AMERICAN SOCIETY FOR TESTING AND MATERIALS (1995)
Standard Guide for Risk-Based Corrective Action Applied at Petroleum Release Sites,
ASTM E1739-95

AMERICAN SOCIETY FOR TESTING AND MATERIALS (1998)
Standard Provisional Guide for Risk-Based Corrective, ASTM PS-104

ASINK, J and van den BRINK, W (1985)
Contaminated Soil, Proceedings of the First International Conference of Contaminated Soil
Netherlands, Matinus Nijoff

DEPARTMENT OF THE ENVIRONMENT (1995)
A Guide to Risk Assessment and Risk Management for Environmental Protection,
HMSO, London (being updated)

DETR (1994)
CLR Report No 4 – Sampling Strategies for Contaminated Land,
HMSO, London.

DETR (in preparation)
CLR9 – Contaminants in Soils: Collation of Toxicological Data and Intake Values For Humans
HMSO, London

DETR (in preparation)
CLR10 – The Contaminated Land Exposure Assessment Model (CLEA): Technical Basis and Algorithms,
HMSO, London

DETR (1999, in draft)
CLR11 - Handbook of Model Procedures for the Management of Contaminated Land,
HMSO, London

ENVIRONMENT AGENCY (1999)
Methodology for the Derivation of Remedial Targets for Soil and Groundwater to Protect Water Resources,
R&D Publication 20, Environment Agency (Bristol)

UNITED STATES ENVIRONMENTAL PROTECTION AGENCY (1989)
Risk Assessment Guidance for Superfund (RAGS), Volume 1, Human Health Evaluation Manual,
Office of Emergency and Remedial Response, EPA/540/1-89/002

UNITED STATES ENVIRONMENTAL PROTECTION AGENCY (1989)
Methods for Evaluating the Attainment of Cleanup Standards, Volume 1: Soils and Solid Media,
Statistical Policy Branch, PB89-234959

UNITED STATES ENVIRONMENTAL PROTECTION AGENCY (1996)
Proposed Guidelines for Carcinogen Risk Assessment,
Office of Research and Development, EEPA/600/P-92/D3C

7 Risk communication

7.1 INTRODUCTION

When the current state or future of contaminated land is the subject of decisions, several groups and individuals are likely to have an interest in the outcome of the decision-making process. They will range from those with financial interests, such as lenders and insurers, through regulators to concerned members of the public and the media. Each will have different expectations of the decision-making process and varying grasp of the technical issues involved. Each will have their own perception of the process, the issues and the expected outcome. In many cases these will conflict. Risk communication is the process of communicating the risk-based decisions to all stakeholders in a clear, concise and transparent manner that allows each to understand, and take part in, the process involved so that the most appropriate decisions can be made.

There has been relatively little history of contaminated land risk assessment in the UK, and even less experience of effective communication of risk. This chapter aims to set out why it is important to communicate risk effectively, who to communicate to, when this should be done and offers suggestions about the approach to take.

7.2 WHY IS RISK COMMUNICATION AN ESSENTIAL PART OF THE RISK ASSESSMENT PROCESS?

Decisions on the future of contaminated land rarely affect just the landowner. Usually there is some regulatory involvement. Commonly, such sites have been contaminated by an industrial or commercial process and are located near a community, which collectively may have concerns about the land. These concerns can encourage a community to bring pressure to bear on regulators or planners and in some circumstances communities can take matters into their own hands to influence the decision-making process. In practical terms, some consequences of this concern for the construction industry can include:

- withdrawal of goodwill to a proposed scheme by the community in which the site is located, resulting in pressure being applied to the local authority at planning stage to reject it or to accept only a substantially modified version of a development planning application

- imposition of conditions to granting of planning consent (for example Section 106 agreements), conferring additional duties on the developer to assess or control contamination to the regulatory authorities' satisfaction

- extreme opposition to plans for the site manifested in actions by individuals or pressure groups that may result in the developer being unable to enter the site or to continue work there.

In these examples the effect may be to delay the programme and thereby to increase costs. There will be times when these are responses to situations that could have been avoided by careful and fully inclusive risk communication. Effective risk communication encourages discussions between the various stakeholders to help reach a common understanding of the situation and the selection of the risk management decision most acceptable to all parties. The risk communication process is enhanced when the risk assessment is used to provide options for management prior to communication to stakeholders showing how contamination can be tackled and made safe.

7.3 WHAT ARE THE DIFFERENT PERCEPTIONS OF RISK?

To understand how to communicate contaminated land risk, it is important to appreciate how risk is perceived by stakeholders in contaminated land.

The term "stakeholder" is the collective term for an interest group (see Figure 1.1), but of course each stakeholder group is made up of individuals. Risks are perceived differently between individuals. An individual's perception of risk will be influenced by their personal values, experience of life, culture and background. The risk assessor should take these into account when communicating risk.

Risks are frequently perceived differently between those whose technical knowledge of a hazardous situation is considerable and those with non-specialist knowledge of the subject. Research has been carried out to determine why there is frequently discrepancy between hazards that experts consider "risky" (that is risks of concern) and those that the general public consider "risky" (Williams 1998). As an illustration, consider the comparative assessment of the following situations, by "experts" and by the public:

Table 7.1 *Examples of perception of the public compared with that of experts (after Williams, 1998)*

Situation	Risks perceived by Experts	Risks perceived by Public
Nuclear power/waste	A moderate risk that is acceptable	An extreme risk that is unacceptable
Radon in homes	A moderate risk that merits positive action	A very low risk met with apathy
Irradiation of food	A low risk that is acceptable	A moderate risk whose acceptability is open to question.
Electric/magnetic fields	A low risk that is acceptable	A risk that is generating increasing concern and becoming unacceptable.

This is a simplification of risk perception since public reaction to a particular hazard issue is by no means uniform. Response to the issue might be driven by concerns such as movement of contaminated material on roads, mobilisation of dust to the atmosphere or non-technical concerns such as loss of amenity areas. Adverse press coverage may also influence the response. The practical effect may be united opposition to particular decisions made about contaminated land by other stakeholders.

Perception of risk is affected by the individual's sense of control over the situation. This is important, since individuals and stakeholder groups are more likely to accept risk if it appears to be within their control. Drivers on the road and mountaineers are both at risk, but driving and mountaineering activities continue nonetheless. By contrast, land contamination, like nuclear risks, is perceived as negative since the potential health impacts are uncertain, long-term, imposed involuntarily and associated with "dreaded" outcomes (Board and O'Connor, 1992). The public tend to view risks associated with involuntary risky activities as unacceptably high, since these may be perceived as having introduced factors to an individual's lifestyle that are beyond his/her control.

Much of the perception of risk is to do with perceived benefit. The apparently "acceptable" risks referred to above (voluntary risks) are regarded as positive, because the individuals involved consider the risk worthwhile in the pursuit of a defined goal –

in this case driving to a destination, or reaching the top of a mountain peak. In contrast, the siting of a nuclear facility or waste disposal site adjacent to an individual's residence is often perceived as introducing hazards that are potentially catastrophic, fatal, delayed and not off-set by any perceived benefits (Petts, 1994).

Research has shown that risks are often perceived on the basis of the individual's familiarity with the situation as well as the sense of control over the situation. Some common practices, such as smoking are widely considered hazardous to health. Smoking involves a high degree of the individual's control over the situation. Where there is more personal choice over the situation the less fear is felt and thus the lesser the perceived risk.

So what are the factors affecting acceptance of risks? One of the most important influences on the general acceptance of risk by stakeholders is the issue of *trust* and *credulity*. If trust is lacking, no form or process of communication will be satisfactory (Slovic 1993). In the current context this refers to trust between the risk communicator and the stakeholders. "Who" is communicating is as important as "what" is being communicated (Petts 1994). The environmental record of the communicator is often crucial, with certain groups and organisations' track record weighing against them. Opposition to contamination management decisions often stems from perceived lack of vigour in the assessment of environmental impact. Wynne (1992) provides illustration of this with his study of the public perception of the effects of the Chernobyl. He examined the relationship between hill farmers of the affected regions of the UK's Lake District and scientists and experts who were advising on the contamination of land and livestock by radioactive fallout. He found that public understanding of science, and therefore risk perception, was not so much about public capabilities in understanding technical information, but about the trust and credibility they invest in scientific experts. In this particular situation, scientific advice was seen to change drastically within a short time of the initial contamination. In addition, scientists were seen to produce theories and undertake assessments and experiments that did not take into account farmers' specialised knowledge of the land and stock and how this might affect the outcome. Essentially, outside experts did not recognise the value of the farmers' own expertise, or consider it necessary to integrate the knowledge with science in order to assess the situation. Farmers observed first-hand inconsistencies in the taking of environmental measurements; where and how samples were taken and observed variability in readings over small distances. This related to "conspiracy theories", reluctance of some organisations to release information and other factors served to undermine the confidence that this stakeholder group had in the scientific assessment advice offered.

From this some general criteria by which lay persons judge scientific risks has been formed (Table 7.2).

Table 7.2 *Lay criteria for judgement of science (adapted from Wynne, 1992)*

Criterion	Comment
Does the scientific knowledge work?	Eg, predictions fail.
Do scientific *claims* pay attention to other knowledge available?	Eg, do the field tests take into account local knowledge about the land?
Does scientific *practice* pay attention to other available knowledge?	Eg, are experiments carried out in a way that other people believe will not work?
Is the form of the knowledge as well as the content recognisable?	Eg, degrees of expressed certainty, aggregation, standardisation.
Are scientists open to criticism?	Eg, do scientists recognise other knowledge and expertise within stakeholder groups?
What are the social and institutional affiliations of experts?	Eg, imputed social and political biases and interests; historical track record of openness, trustworthiness.
What issue "overspill" exists in lay experience?	Eg, what perception of related risky activities is connected with other stakeholders or with the activity itself, encouraging pre-formed opinion?

7.4 HOW TO COMMUNICATE CONTAMINATED LAND RISKS

The process of communication risk might seem straightforward, since communication of ideas and other interaction between individuals and groups takes place daily. However, the communication becomes complicated by the wide diversity of interest within the audience, uncertainty, and sometimes competing messages, which make this a more challenging task.

Historically, risk communication has been seen as a process undertaken to legitimise decision-making rather than to enhance it (Petts, 1999). There is increasing recognition that communication should involve all stakeholders, and at appropriate times within the assessment process. This section describes ways of identifying groups with which communication will take place, when communication occurs and how this may be done.

Risk communication is an acquired skill and assessors should recognise their own limitations in this area and seek appropriate training or assistance, as required.

There are four stages involved in communicating risks.
1. Determine who to communicate with.
2. Choose the time to communicate risk.
3. Decide what to communicate.
4. Decide how to communicate.

At the outset, it is helpful to have some idea of what the communication is designed to achieve. The communicator should have, in advance, some idea of possible outcomes of the process, particularly those where there is likely to be mutual understanding and agreement, and work towards these.

Step 1 – determining WHO to communicate with

At the start of the risk assessment process, it is essential for the assessor to identify the people with whom they will need to communicate. In general, these are the stakeholders

(Chapter 1), who have an interest in the contaminated land for whatever reason. These stakeholders have differing technical expertise, knowledge of the site and expectations.

Particular care needs to be taken when identifying local community stakeholders. Whereas professional stakeholders such as landowners and funders are relatively easy to identify, it is harder to identify individuals with an interest. The latter may not necessarily live next to the site. They may dwell some distance away but consider themselves at risk from migration of contamination off site (such as living along a main road along which site construction traffic will pass) or due to concerns prompted by national publicity (airborne contamination). When identifying stakeholders the assessor should be aware of the "public relations history" of the site. Some sites may have been subject to one or more abandoned attempts at redevelopment, where environmental issues were handled badly. As a result of this experience, feelings of distrust of the assessment process will tend to be enhanced in the minds of many stakeholders. The assessor will need to be aware of previous failures to tackle real or perceived contamination problems, which may be linked with perceptions of local blight.

Sometimes the stakeholders closest to the site in question will be the least concerned about the outcome of the assessment process and it may be difficult to establish communication with these. This may be due to the perceived familiarity and "control" over the situation discussed earlier. Consider the example in Box 7.1:

Box 7.1 *Example of community concerns over a contaminated site*

> A large multinational company developed a new office building on a 30-year-old landfill site. The works involved excavating and exposing some of the wastes. Regular meetings held with local residents to discuss environmental concerns were attended by residents of a village community some 500 m away, whereas many of those actually living immediately adjacent to the site did not attend and did not express any concerns when formally approached.

Step 2 – determining WHEN to communicate

Although this discussion of risk communication appears at the end of this guidance, communication is an ongoing process, paralleling each of the steps of the assessment. There will of course be stages of the assessment in which technical uncertainties have been minimised and the outcomes of the assessment can be communicated with greater confidence. It is important not to begin risk communication too late in the process. Consider the example in Box 7.2.

Box 7.2 *Example of mistiming the risk communication process*

> This example concerns a former tip, that later became a park with full public access. A preliminary study of the site identified that drums of chemical waste had been dumped there some 30 years previously, before the site had been turned into parkland. The owner of the site decided to carry out an intrusive site investigation. The site investigation company contracted to carry out the investigation prepared a health and safety plan for its staff that identified the need for protective clothing to be worn during the site works. Park users first became aware that there was a potential problem when they saw among them scientists wearing white contamination protection suits and breathing apparatus taking samples of soil and groundwater. This prompted a flurry of speculation, particularly in the local media, and enquiries to local authority officers, regulators and local representatives from park users and local residents expressing fears about potential hazards to health. Soon use of the park all but ceased while local authority staff attempted to allay this local concern.

When risks are communicated late in the assessment programme, some stakeholders may feel that the process has been decided without their being able to influence the outcome. In consequence, these stakeholders may be less likely to accept the outcome.

Step 3 – determining WHAT to communicate

An important objective of risk communication is allow stakeholders to understand the risks, how it will affect them and to allow them to have a stake in the decision-making process. The assessor must decide on the key messages to communicate to stakeholders.

For schemes where a wide variety of parties is involved, the communication of risk should be tailored to the specific receiving group. The assessor might consider producing a report for two or more levels of understanding. Other disciplines have adopted this approach. An example is environmental impact assessment, where, for public schemes in particular, it is usual to produce a technical report of the assessment, accompanied by a non-technical summary for members of the public.

Step 4 – deciding HOW to communicate

This involves decisions on the manner in which the message is delivered and also the circumstances during which the communication occurs.

Risk communication should be the process of exchange of information on risk assessment and risk-based decisions between all of the stakeholders in contaminated land. One of the problems of risk communication is that it is often seen as a process that passes information from the expert assessor to other interested parties. In particular, the outcome of a risk assessment is often used to educate non-technical parties, such as members of the public. However, the process of educating non-technical stakeholders can increase the sense that their control over the hazardous activity is minimal.

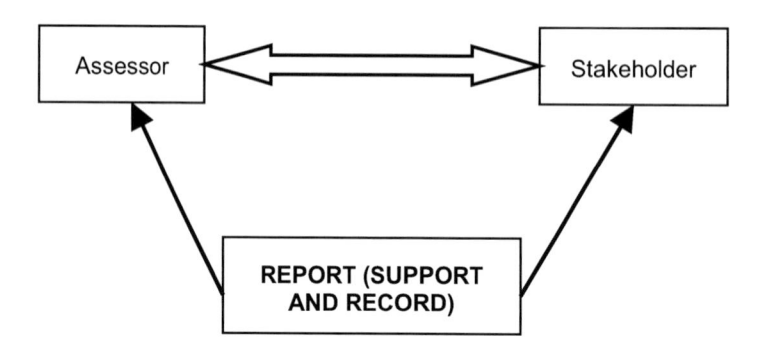

Different stakeholders will have different abilities and levels of understanding of technical issues. However, all technical assessment processes can be communicated intelligibly if care is taken in the explanation. If risk communicators are perceived as reserving or holding back information in any way, the risk process will not be seen as clear and transparent, so the outcome is less likely to be accepted by all parties.

7.5 COMMUNICATING UNCERTAINTIES

It is important to be aware of the uncertainties affecting the assessment and to communicate these so that the reasonableness of the risk estimation can be justified to other stakeholders.

It is important to distinguish between uncertainty and ignorance. The latter arises from the lack of awareness of matters that influence the assessment. For example, ignorance of the circumstances at the site might result in not all of the hazards present being properly identified, or all of the pathways evaluated. Uncertainty, by contrast, "is a state of knowledge in that, although the factors influencing the issue are identified, their effects cannot be precisely described" (HSE, 1999). In other words, uncertainty is the state of understanding that arises from the gaps in knowledge about one or more areas of the risk assessment.

In practical terms, uncertainty arises from information gaps that are associated with the data available about a site. Uncertainties commonly arise from:

- lack of knowledge about the characteristics of the contaminants, pathways and receptors
- variations in the exposure to hazards by receptors, and in the dose-response relationship
- limited data of the effects of hazards on receptors (such as limited availability of data on the environmental fate of chemicals)
- the quality of sampling and analysis data.

The assessor should identify all the gaps in their knowledge and the assumptions made to overcome these. This is an essential part of the transparency of the assessment process. Where it is appropriate, the assessor should evaluate the effect on the outcome of the assessment that would arise if the assumption were to be modified. The assessor should decide whether the amount of uncertainty allowed in the assessment is reasonable. Eventually the point is reached where the uncertainties and lack of understanding are such that the assessment must be based largely or wholly on expert opinion. This expert judgement is an inherent part of the risk assessment process, but the assumptions made should be clearly explained and made transparent. The quality of the assessment then depends on such characteristics as credulity, standing and independence of the assessors (McQuaid J, 1999).

7.6 COMMUNICATION OF RISK TO OTHER ASSESSORS AND REMEDIAL CONTRACTORS

Although the assessment of risk forms part of the overall risk management process, in practice the assessor may not be the person who makes risk based decisions or carries them through to remedial action. In addition it has been common industry practice for a different assessor to be involved at different points in the process. This is because the risk assessment process is often seen as a series of discrete entities – desk study, prioritisation, site investigation, etc., effectively compartmentalising the process, and in some cases the individual or organisation responsible for the overall risk management process appoints assessors for each successive stage of the assessment. The likelihood of this occurring is increased where decisions on risk are deferred, or where there is a lengthy delay between each stage of the assessment, due perhaps to site access difficulties or obtaining funding. A good assessment should avoid compartmentalisation and the steps should be continuous, sometimes overlapping, each step distinguishable only because it involves a successive refinement of data. If different assessors are appointed to carry out individual assessment tasks, it is essential that all these communicate with each other, to ensure that all data, findings and supporting information are made available for successive steps in the process. This will save not only time, effort and budget spent duplicating the gathering of data, but also will ensure that important information is not overlooked. A thorough report with all supporting information correctly included or referenced will assist this.

When the assessment is complete and the communication process with stakeholders completed, there should be a mechanism by which risks can be communicated to new parties that may take an interest in the site at some point in the future. For the construction industry, this means ensuring that risks about a site are communicated to development contractors or those who may come into contact with contamination in the course of any future works at the site. Where a site is proposed for redevelopment, the results of the risk assessment should be made available, either supplied with tender or contract documentation or else made available for inspection by contractors and other parties. This will enable contractors and designers to carry out statutory commitments with respect to health and safety planning (COSHH, CDM etc) and to prepare remedial designs and site management plans, as required. The CDM planning supervisor should ensure that this happens.

7.7 CONCLUSION

Risk perception will greatly influence the type or level of risk considered "acceptable". The perception of risk by all the interested parties may not be the same as the level of risk assumed by those undertaking the assessment. Public opinion can greatly influence decisions. For this reason, risk communication should be made an integral part of the risk management process. Successful risk communication requires the involvement of the stakeholders in the decision making process.

Box 7.3 indicates some tips for good practice.

Box 7.3 *Tips for good practice*

Always be prepared to communicate risk at all stages at the assessment process.

Allow for different approaches to communication for different stakeholders.

Allow the process to be clear and consistent. At all stages consideration should be given to ensuring that actions and communications are honest, transparent, competent and credible.

Do not withhold information, but limit the communication to a few clear messages to explain the situation. Ensure that the communication is helpful rather than reactionary and defensive.

Admit uncertainties in the estimation and evaluation and gaps in the data.

Be prepared for situations where the contaminants in question are particularly emotive. Examples include radioactivity and asbestos. Contaminants such as these are well known in the minds of most stakeholders, who are likely to be less accepting of "solutions" involving such substances. In this case openness is especially crucial.

Be careful in the use of jargon. The words "contaminated", "land" and "harm" in the same sentence are emotive enough, but the incautious use of dramatic phrases such as "explosion hazard" or "risk of toxic effects" is unwise. Similarly beware of descriptive terminology. For example phrases such as "not very significant" are open to interpretation particularly by those without extensive technical training, and stakeholders may interpret this in a different way from the assessor.

Make communication a two-way process. The assessor should be prepared to listen as well as to instruct.

The communicator should be available to answer stakeholders' concerns as much as possible. A communicator who can be relied upon to attend meetings, replies to all correspondence, and is seen to listen to genuine concerns is more likely to gain the trust of interest groups and individuals, forming the basis of mutual trust.

It is useful for the assessor to give some examples of situations where contamination has been successfully managed.

7.8 FURTHER READING

HEALTH & SAFETY EXECUTIVE (1999)
Reducing Risks, Protecting People, HSE discussion document
Health & Safety Executive, London

ILGRA (1998)
Risk Communication, a Guide to Regulatory Practice
Risk Assessment Policy Unit, Health and Safety Executive, London

McQUAID, J (1999).
Regulatory Decision Making on Risk, Paper to Seminar to Centre for Environmental Strategy, University of Surrey
Health and Safety Executive, London.

PETTS, J (1999)
"Public Participation in EIA", in *Handbook of Environmental Impact Assessment*, J Petts (ed), vol 1, 145–177
Blackwell Science, Oxford, UK

ROYAL SOCIETY (1992)
Risk: Analysis, Perception and Management, Report of a Royal Society Study Group
Royal Society, London

SCOTLAND & NORTHERN IRELAND FORUM FOR ENVIRONMENTAL RESEARCH (SNIFFER, 1999)
Communicating Understanding of Contaminated Land Risks
Scotland & Northern Ireland Forum for Environmental Research

References

AMERICAN SOCIETY FOR TESTING AND MATERIALS (1995)
Standard Guide for Risk-Based Corrective Action Applied at Petroleum Release Sites
ASTM, E1739-95

AMERICAN SOCIETY FOR TESTING AND MATERIALS (1998)
Standard Provisional Guide for Risk-Based Corrective Action
ASTM PS-104

BORD, R and O'CONNOR, R (1992)
"Determinants of Risk Perception of a Hazardous Waste Site"
Risk Analysis, vol 12, no 3, pp 41–46

BRE (1996)
Sulphate and acid resistance of concrete in the ground, BRE Digest 363
Building Research Establishment, Watford, UK

BRITISH DRILLING ASOCIATION (1991)
Guidance Notes for the Safe Drilling of Landfill, contaminated land and adjacent areas
BDA Brentwood, UK

BRITISH STANDARDS INSTITUTION (1987)
BS 6068-6.4 *Water Quality. Sampling. Guidance on sampling from lakes, natural and man-made*
British Standards Institution, London

BRITISH STANDARDS INSTITUTION (1991)
BS 6068-6.6 *Water Quality. Sampling. Guidance on sampling from rivers and streams*
British Standards Institution, London

BRITISH STANDARDS INSTITUTION (1999)
BS 5930 (1999) *Code of Practice for Site Investigation*
British Standards Institution, London

BRITISH STANDARDS INSTITUTION (2000 in draft)
Revision of BS DD175 Investigation of Potentially Contaminated Sites – Code of Practice, 3rd draft
British Standards Institution, London

CBI (1993)
Firm Foundations – CBI Proposals for Environmental Liability and Contaminated Land
Confederation of British Industry, London

CIRIA (1996)
A Guide for Safe Working Practices on Contaminated Sites, Report 132
CIRIA, London

CONCERTED ACTION ON RISK ASSESSMENT IN THE EUROPEAN UNION (CARACAS) (1999)
Risk Assessment for Contaminated Sites in Europe, Volume 1: Scientific Basis
LQM Press, Nottingham

CROWHURST, D and MANCHESTER, S J (1993)
The Measurement of Methane and Other Gases from the Ground, Report 131
CIRIA, London

DEPARTMENT OF THE ENVIRONMENT (1994)
Guidance on preliminary site inspection of contaminated land, Volumes 1 and 2, CLR 2, report by Applied Environmental Research Centre Ltd
HMSO, London

DEPARTMENT OF THE ENVIRONMENT (1994)
Documentary research on industrial sites, CLR 3, report by RPS Group plc
HMSO, London

DEPARTMENT OF THE ENVIRONMENT (1995)
A Guide to Risk Assessment and Risk Management for Environmental Protection
HMSO, London (being updated)

DEPARTMENT OF THE ENVIRONMENT (1996)
Draft Statutory Guidance on Contaminated Land (NB: more recent versions, 1999, available from DETR)
HSMO, London

DETR (1994)
Sampling Strategies for Contaminated Land, CLR 4
HMSO, London

DETR (1998)
Control and Remediation of Radioactively Contaminated Land – A Consultation Paper
DETR, London

DETR (2000)
Draft *Circular on Contaminated Land*
DETR, London

DETR (in preparation)
Contaminants in Soils: Collation of Toxicological Data and Intake Values for Humans, CLR 9
HMSO, London

DETR (in preparation)
The Contaminated Land Exposure Assessment Model (CLEA): Technical Basis and Alogorithms, CLR 10
HMSO, London

DETR (1999, in draft)
Handbook of Model Procedures for the Management of Contaminated Land, CLR 11
HMSO, London

DETR (in preparation)
Acceptance Criteria for Contaminated Soils
HMSO, London

ENVIRONMENT AGENCY (1999)
Methodology for the Derivation of Remedial Targets for Soil and Groundwater to Protect Water Resources, R&D Publication 20
Environment Agency, Bristol

ENVIRONMENT AGENCY (1999)
The Environment Agency and land contamination
Environment Agency, Bristol

ENVIRONMENT AGENCY (in preparation)
Development of Soil Sampling Strategies for Contaminated Land
Environment Agency, Bristol

ENVIRONMENT AGENCY (in preparation)
Guidance on short-term risks to human health, Project P5-039

ENVIRONMENT AGENCY (in preparation)
Guidance on site-specific assessment of chronic risks to human health from contamination, Project P5-041

ENVIRONMENT AGENCY (in preparation)
Guidance on the risks of contaminated land to buildings, building materials and services, Project P5-035

ENVIRONMENT AGENCY (in preparation)
Collation of toxicological data and development of guideline values for explosive substances

FERGUSON, C and ABBACHI, A (1993)
"Incorporating Expert Judgement into Statistical Designs for Contaminated Sites"
Land Contamination & Reclamation, vol 1, no 3

THE FORUM ON CONTAMINATION IN LAND (1999)
Standard Forms for Requesting Laboratory Services and for the Laboratory Response
FOCIL, UK

HARRIS, M R, HERBERT, S M and SMITH, M A (1995)
Remedial Treatment for Contaminated Land, Volume III: Site Investigation and Assessment, Special Publication 103
CIRIA, London

HEALTH AND SAFETY EXECUTIVE (1991)
Protection of Workers and the General Public during Development of Contaminated Land, HS(G)66
HMSO, London

HEALTH & SAFETY EXECUTIVE (1999)
Reducing Risks, Protecting People, HSE discussion document
Health & Safety Executive, London

HEALTH AND SAFETY EXECUTICE (2000)
Occupational exposure limits 2000, EH40
HSE Books, Sudbury

ILGRA (1998)
Risk Communication, A Guide to Regulatory Practice
Risk Assessment Policy Unit, Health and Safety Executive, London

KELLY, R T (1980)
"Site Investigation and Material Problems", in *Proceedings of a conference on the reclamation of contaminated land*, pp B2/1–B2/14
Society of Chemical Industry, London

McQUAID, J (1999)
Regulatory Decision Making on Risk, Paper at Seminar to Centre for Environmental Strategy, University of Surrey
Health and Safety Executive, London

MHSPE (Ministry of Housing Spacial Planning and Environment, Holland) (1994)
Circular on Intervention Values for Soil Remediation
Government Printing Office, The Hague, Netherlands

NATIONAL HOUSE-BUILDING COUNCIL (1998)
"Land Quality – Managing Ground Conditions" (Chapter 4.1) in *Standards*
NHBC, UK

ONIONS, K R, WHITWORTH, K and JACKSON, P (1996)
"Application of Geophysical methods to site investigations at contaminated old collieries"
Quarterly Journal of Engineering Geology, 29, pp 219–231
The Geological Society, UK

PETTS, J (1994)
"Effective Waste Management: Understanding and Dealing with Public Concerns"
Waste Management & Research, 12, pp 207–222)

PETTS, J (1999)
"Public Participation in EIA", in J Petts (ed) *Handbook of Environmental Impact Assessment*, vol 1, pp 145–177
Blackwell Science, Oxford, UK

ROYAL SOCIETY (1992)
Risk: Analysis, Perception and Management, Report of a Royal Society Study Group
Royal Society, London

SCOTLAND & NORTHERN IRELAND FORUM FOR ENVIRONMENTAL RESEARCH (SNIFFER, 1999)
Communicating Understanding of Contaminated Land Risks
Scotland & Northern Ireland Forum for Environmental Research

SCOTTISH ENTERPRISE (1998)
How to Investigate Contaminated Land, requirements for Contaminated Land Site Investigations, 2nd edition
Scottish Enterprise, Glasgow

SLOVIC (1993)
"Perceived Risk, Trust and Democracy"
Risk Analysis, vol 13, no 6

UNITED STATES ENVIRONMENTAL PROTECTION AGENCY (1989)
Risk Assessment Guidance for Superfund (RAGS), Volume 1: Human Health Evaluation Manual, EPA/540/1-89/002
Office of Emergency and Remedial Response

UNITED STATES ENVIRONMENTAL PROTECTION AGENCY (1989)
Methods for Evaluating the Attainment of Cleanup Standards, Volume 1: Soils and Solid Media, PB89-234959
Statistical Policy Branch

VAN DEN BERG, R, DENNEMAN, C A J and ROELS, J M (1993)
"Risk Assessment of Contaminated Soil: Proposals for Adjusted, Toxicologically Based Dutch Soil Clean-up Criteria", in F Arendt, G J Annokkee, R Bosman, W J van den Brink (eds), *Contaminated Soil* 93, pp 349–364
Kluwer Academic Publishers, Netherlands

WELSH DEVELOPMENT AGENCY (1993)
The WDA Manual on the Remediation of Contaminated Land
ECOTEC & Environmental Advisory Unit Ltd

WILLIAMS, D R (1998)
What is Safe? The Risks of Living in a Nuclear Age
Royal Society of Chemistry, London

WYNNE, B (1992)
"Misunderstood misunderstanding: Social identities and public uptake of science"
Public Understanding of Science, 1, pp 281–304

A1 Publications on contaminated land for further reading

A1.1 UK GOVERNMENT PUBLICATIONS – GENERAL PUBLICATIONS

DEPARTMENT OF THE ENVIRONMENT (1995)
A Guide to Risk Assessment and Risk Management for Environmental Protection
HMSO, London

DEPARTMENT OF THE ENVIRONMENT and WELSH OFFICE (1987)
Development of Contaminated Land, Circulars 21/87 and 22/87 (Welsh Office)
(superseded by PPG23, as it relates to England)
HMSO, London

DEPARTMENT OF THE ENVIRONMENT and WELSH OFFICE (1989)
Landfill Sites: Development Control, Circulars 17/89 (DoE) and 38/89 (Welsh Office)
HMSO, London

DEPARTMENT OF THE ENVIRONMENT (1994)
Planning Policy Guidance: Planning and Pollution Control, PPG23
HMSO, London

DEPARTMENT OF THE ENVIRONMENT (1994)
Framework for Contaminated Land: Outcome of the Government's Policy Review and Conclusions from the Consultation Paper Paying for our Past
HMSO, London

DETR (2000)
Environmental Protection Act 1990: Part IIA Contaminated Land, Circular 02/2000
HMSO, London

DETR (2000)
Environmental Protection, England, the Contaminated Land (England) Regulations 2000, Statutory Instrument 2000 No 227
HMSO, London

A1.2 REPORTS SPONSORED BY UK DEPARTMENT OF ENVIRONMENT, TRANSPORT AND THE REGIONS (DETR) AND OTHER GOVERNMENT DEPARTMENTS

A1.2.1 Contaminated land workshops

DEPARTMENT OF THE ENVIRONMENT (1995)
Workshop 1, Professional Standards, April 1994
HMSO, London

DEPARTMENT OF THE ENVIRONMENT (1995)
Workshop 2, Insurance Standards, July 1994
HMSO, London

DEPARTMENT OF THE ENVIRONMENT (1995)
Workshop 3, Risk assessment and the use of guidelines, July 1994
HMSO London

DEPARTMENT OF THE ENVIRONMENT (1995)
Workshop 4, Analytical methods for soils from contaminated land, October 1994
HMSO, London

DEPARTMENT OF THE ENVIRONMENT (1995)
Workshop 5, Specifications, contracts, insurances and warranties, February 1995
HMSO, London

A1.2.2 ICRCL (Inter-Departmental Committee on the Redevelopment of Contaminated land) publications

DEPARTMENT OF THE ENVIRONMENT (1983)
Notes on the redevelopment of sewage works and farms, ICRCL Guidance Note 23/79, 2nd edition, November 1983
HMSO, London

DEPARTMENT OF THE ENVIRONMENT (1986)
Notes on the redevelopment of gasworks sites, ICRCL Guidance Note 18/79, 5th edition, April 1986
HMSO, London

DEPARTMENT OF THE ENVIRONMENT (1986)
Notes on the fire hazards of contaminated land, ICRCL Guidance Note 61/84, 2nd edition, July 1986
HMSO, London

DEPARTMENT OF THE ENVIRONMENT (1987)
Guidance on the assessment and redevelopment of contaminated land, ICRCL Guidance Note 59/83, 2nd edition, July 1987
HMSO, London

DEPARTMENT OF THE ENVIRONMENT (1987)
Notes on the redevelopment of scrap yards and similar sites, ICRCL Guidance Note 42/80, 2nd edition, October 1983
HMSO, London

DEPARTMENT OF THE ENVIRONMENT (1990)
Notes on the restoration and aftercare of metalliferous mining sites for pasture and grazing, ICRCL Guidance Note 70/90, 1st edition, February 1990
HMSO, London

DEPARTMENT OF THE ENVIRONMENT (1990)
Asbestos on contaminated sites, ICRCL Guidance Note 64/85, 2nd edition, October 1990
HMSO, London

DEPARTMENT OF THE ENVIRONMENT (1990)
Notes on the development and after-use of landfill sites, ICRCL Guidance Note 17/78, 8th edition, December 1990
HMSO, London

A1.2.3 CLR published research reports

DEPARTMENT OF THE ENVIRONMENT (1994)
A framework for assessing the impact of contaminated land on groundwater and surface water, Volumes 1 and 2, CLR 1
Report by Aspinwall & Co, HMSO, London

DEPARTMENT OF THE ENVIRONMENT (1994)
Guidance on preliminary site inspection of contaminated land, Volumes 1 and 2, CLR 2
Report by Applied Environmental Research Centre Ltd, HMSO, London

DEPARTMENT OF THE ENVIRONMENT (1994)
Documentary research on industrial sites, CLR 3
Report by RPS Group plc, HMSO, London

DEPARTMENT OF THE ENVIRONMENT (1994)
Sampling strategies for contaminated land, CLR 4
Report by Centre for Research into the Built Environment, Nottingham Trent University, HMSO, London

DEPARTMENT OF THE ENVIRONMENT (1994)
Information systems for land contamination, CLR5
Report by Meta Generics Ltd, HMSO, London

DEPARTMENT OF THE ENVIRONMENT (1995)
Prioritisation & categorisation procedure for sites which may be contaminated, CLR6
Report by M J Carter Associates, HMSO, London

DEPARTMENT OF THE ENVIRONMENT (1997)
A Quality Approach for Contaminated Land Consultancy, CLR8
Report by the Environmental Industries Commission in Association with the Laboratory of the Government Chemist, HMSO, London

A1.2.4 CLR research reports in preparation

DETR (1999, in draft)
Overview of guidance on the Assessment of Contaminated Land, CLR7
Report by Chrisalis Environmental Consulting

DETR (1999, in draft)
Potential Contaminants for the Assessment of Land, CLR8
Prepared by CES

DETR (1999, in draft)
Contaminants in Soils: collation of toxicological data and intake values for humans, Consolidated Main Report, CLR9
Prepared by RPS Consultants Ltd

DETR (1999, in draft)
Contaminant in Soils: collation of toxicological data and intake values for humans (various contaminants), CLR9 TOX 1-10
Prepared by RPS Consultants Ltd

DETR (1999, in preparation)
Handbook of the Model Procedures for the Management of Contaminated Land, CLR11
Prepared by the Environment Agency, Bristol

DETR (1999, in preparation)
Guideline Values for Contamination in Soils, CLR10 GV1-10
Prepared by the Centre for Research into the Built Environment, Nottingham Trent University

DEPARTMENT OF THE ENVIRONMENT (2000, in draft)
The Contaminated Land Exposure Assessment Model (CLEA): technical basis and algorithms, CLR10
Prepared by the Centre for Research into the Built Environment, Nottingham Trent University

DEPARTMENT OF THE ENVIRONMENT (2000, in draft)
Potential Contaminants for the Assessment of Land, CLR8
Report by Consultants in Environmental Science)

DEPARTMENT OF THE ENVIRONMENT (2000, in draft)
A series of reports is being drafted, as below

> *Guideline values for arsenic contamination in soils*, CLR10 GV1
> Report by the Centre for Research into the Built Environment, Nottingham Trent University
>
> *Guideline values for cadmium contamination in soils*, CLR10 GV2
> Report by the Centre for Research into the Built Environment, Nottingham Trent University
>
> *Guideline values for chromium contamination in soils*, CLR10 GV3
> Report by the Centre for Research into the Built Environment, Nottingham Trent University

Guideline values for contamination in soils by inorganic cyanides, CLR10 GV4
Report by the Centre for Research into the Built Environment, Nottingham Trent University

Guideline values for lead contamination in soils, CLR10 GV5
Report by the Centre for Research into the Built Environment, Nottingham Trent University

Guideline values for inorganic mercury contamination in soils, CLR10 GV6
Report by the Centre for Research into the Built Environment, Nottingham Trent University

Guideline values for nickel contamination in soils, CLR10 GV7
Report by the Centre for Research into the Built Environment, Nottingham Trent University

Guideline values for phenol contamination in soils, CLR10 GV8
Report by the Centre for Research into the Built Environment, Nottingham Trent University

Guideline values for benzo(a)pyrene contamination in soils, with advice on assessing human health risks from mixtures of polycyclic aromatic hydrocarbons, CLR10 GV9
Report by the Centre for Research into the Built Environment, Nottingham Trent University

Guideline values for selenium contamination in soils, CLR10 GV10
Report by Centre for Research into the Built Environment, Nottingham Trent University

A1.2.5 Industry profiles

The DETR (formerly DoE) Industry Profiles provide developers, local authorities and anyone else interested in contaminated land, with information on the processes, materials and wastes associated with individual industries. They also provide information on the contamination which might be associated with specific industries, factors that affect the likely presence of contamination, the effect of mobility of contaminants and guidance on potential contaminants. They are not definitive studies but they introduce some of the technical considerations that need to be borne in mind at the start of an investigation for possible contamination. These include:

- Airports
- Animal and animal products processing works
- Asbestos manufacturing works
- Ceramics, cement and asphalt manufacturing works
- Chemical works: coatings (paints and printing inks) manufacturing works
- Chemical works: explosives, propellants and pyrotechnics manufacturing works
- Chemical works: fertiliser manufacturing works
- Chemical works: fine chemicals manufacturing works
- Chemical works: inorganic chemicals manufacturing works
- Chemical works: linoleum, vinyl and bitumen-based floor covering manufacturing works
- Chemical works: mastics, sealants, adhesives and roofing felt manufacturing works
- Chemical works: pesticides manufacturing works
- Chemical works: pharmaceuticals manufacturing work

- Chemical works: rubber processing works (including works manufacturing tyres or other rubber products)
- Chemical works: soap and detergent manufacturing works
- Dockyards and dockland
- Engineering works: aircraft manufacturing works
- Engineering works: electrical and electronic equipment manufacturing works (including works manufacturing equipment containing PCBs)
- Engineering works: mechanical and ordnance works
- Engineering works: railway engineering works
- Engineering works: shipbuilding, repair and shipbreaking (including naval shipyards)
- Engineering works: vehicle manufacturing works
- Gasworks, coke works and other coal carbonisation plants
- Metal manufacturing, refining and finishing works: electroplating and other metal finishing works
- Metal manufacturing, refining and finishing works: precious metal recovery works
- Oil refineries and bulk storage of crude oil and petroleum products
- Power stations (excluding nuclear power stations)
- Pulp and paper manufacturing works
- Railway land
- Road vehicle fuelling, service and repair: garages and filling stations
- Road vehicle fuelling, service and repair: transport and haulage centres
- Sewage works and sewage farms
- Textile works and dye works
- Timber products manufacturing works
- Waste recycling, treatment and disposal sites: drum and tank cleaning and recycling plants
- Waste recycling, treatment and disposal sites: hazardous waste treatment plants
- Waste recycling, treatment and disposal sites: landfills and other waste treatment or waste disposal sites
- Waste recycling, treatment and disposal sites: metal recycling sites
- Profile of miscellaneous industries, incorporating:
  - Charcoal works
  - Dry-cleaners
  - Fibreglass and fibreglass resins manufacturing works
  - Glass manufacturing works
  - Photographic processing industry
  - Printing and bookbinding works

A1.3 ENVIRONMENT AGENCY PUBLICATIONS

The Environment Agency has produced various publications relevant to contaminated land risk assessment. Several of these publications are still being drafted and were not available at the time of publication of this guide.

ENVIRONMENT AGENCY (1998)
Policy and practice for the protection of groundwater, 2nd edition
Environment Agency, Bristol

ENVIRONMENT AGENCY (1999)
Methodology for the derivation of remedial targets for soil and groundwater to protect water resources, Project P2-087
Report by Aspinwall & Co Ltd, Environment Agency, Bristol

ENVIRONMENT AGENCY (2000)
Costs and benefits associated with the remediation of contaminated groundwater: a framework for assessment, Project P2-078
Environment Agency, Bristol

ENVIRONMENT AGENCY (2000)
Guidance for the safe development of housing on land affected by contamination, R&D Technical Report P66
HMSO, London

ENVIRONMENT AGENCY (2000)
Costs and benefits associated with the remediation of contaminated groundwater: a framework for assessment, Project P2-078
Environment Agency, Bristol

ENVIRONMENT AGENCY (2000)
Technical support materials for the regulation of radioactively contaminated land, R&D Technical Report P307 (P3-055), prepared by ENTEC UK Ltd and National Radiological Protection Board
Environment Agency, Bristol

ENVIRONMENT AGENCY (in preparation)
Guidance on the risks of contaminated land to buildings, building materials and services, Project P5-035

ENVIRONMENT AGENCY (in preparation)
Piling and penetrative improvement methods on land affected by contamination: initial technical guidance on pollution prevention, National Groundwater and Contaminated Land Centre Report NC/99/72

ENVIRONMENT AGENCY (in preparation)
Development of appropriate soil sampling strategies for land contamination, R&D Technical Report HOCO352

ENVIRONMENT AGENCY (in preparation)
Guidance on short-term risks to human health, Project P5-039

ENVIRONMENT AGENCY (in preparation)
Supporting guidance for the verification of risk assessment and risk management measures, Project P5-046

ENVIRONMENT AGENCY (in preparation)
Methodology for comparison of CLEA with other human health risk assessment packages, National Groundwater and Contaminated Land Centre Report NC/06/07

ENVIRONMENT AGENCY (in preparation)
Guidance on site-specific assessment of chronic risks to human health from contamination, Project P5-041

ENVIRONMENT AGENCY (2000, in draft)
Assessing risks to ecosystems from land contamination, R&D Technical Report P338, Draft Report

SNIFFER (Scotland and Northern Ireland Forum for Environmental Research) (2000)
Framework for deriving numeric targets to minimise the adverse human health effects of long-term exposure to contaminants in soil, Report SR99 (02) f

A1.4 BRE (BUILDING RESEARCH ESTABLISHMENT) PUBLICATIONS

BUILDING RESEARCH ESTABLISHMENT (1991)
Construction of new buildings on gas-contaminated land, Report BR212
BRE, Watford, UK

CHARLES, J A (1993)
Building on fill: geotechnical aspects, Report BR230
BRE, Watford, UK

CROWHURST, D, BEEVER, P F
Fire and Explosion Hazards Associated with the Redevelopment of Contaminated Land. Fire Research Station, Information Paper 1987 IP2/87
BRE, Watford, UK

GARVIN, S L, PAUL, V, UBEROI, S (1995)
Polymeric anti-corrosion coatings for protection of materials in contaminated land, Report BR286
BRE, Watford, UK

PAUL, V (1994)
Performance of building materials in contaminated land, Report BR255
BRE, Watford, UK

PAUL, V (1995)
Bibliography of Case Studies on Contaminated Land: Investigation, remediation and redevelopment, Report BR291
BRE, Watford, UK

A1.5 CONSTRUCTION INDUSTRY RESEARCH AND INFORMATION ASSOCIATION (CIRIA) PUBLICATIONS

CARD, G B (1995)
Protecting development from methane, Report 149
CIRIA, London

CROWHURST, D AND MANCHESTER, S J (1993)
The measurement of methane and other gases from the ground, Report 131
CIRIA, London

HARRIES, C R, WITHERINGTON, P J and MCENTEE, J M (1995)
Interpreting measurements of gas in ground, Report 151
CIRIA, London

HARRIS, M R, HERBERT, S M and SMITH M A (1995)
Remedial treatment for contaminated land, Volume III: Site investigation and assessment, Special Publication 103
CIRIA, London

HARTLESS, R (1992)
Methane and associated hazards to construction: A Bibliography, Special Publication 79
CIRIA, London

HOOKER, P J AND BANNON, M P (1993)
Methane: Its occurrence and hazards in construction, Report 130
CIRIA, London

LEACH B A AND GOODGER H K (1991)
Building on Derelict Land, Special Publication 78
CIRIA, London

O'RIORDAN, N and MILLOY, C J (1995)
Risk assessment for methane and other gases from the ground, Report 152
CIRIA, London

RAYBOULD, J G, ROWAN, S P and BARRY, D L (1995)
Methane investigation strategies, Report 150
CIRIA, London

STEEDS, J E, SHEPHERD, E and BARRY, D L (1996)
A guide for safe working practices on contaminated sites, Report 132
CIRIA, London

Available from Construction Industry Research Information Association, 6 Storey's Gate, Westminster, London SW1P 3AU (tel: 020 7222 8891).

A1.6 HEALTH AND SAFETY EXECUTIVE PUBLICATIONS

HEALTH AND SAFETY EXECUTIVE (1991)
Protection of workers and the general public during the development of contaminated land, HS(G)66
Health & Safety Executive, London

HEALTH AND SAFETY EXECUTIVE (1996)
Development and Validation of an Analytical Method to Determine the Amount of Asbestos in Soils and Loose Aggregates
HSE Books, Sudbury, Suffolk

HEALTH AND SAFETY EXECUTIVE (1999)
Reducing Risks, Protecting People, discussion document
HSE Books, Sudbury, Suffolk

A1.7 OTHER PUBLICATIONS

BRITISH STANDARDS INSTITUTION (1999)
BS 5930 (1999) *Code of Practice for Site Investigation*
British Standards Institution, London

BRITISH STANDARDS INSTITUTION (2000 in draft)
Revision of BS DD175 Investigation of potentially contaminated sites – code of practice
3rd draft
British Standards Institution, London

HIGHWAYS AGENCY (1995)
Site investigation for highway works on contaminated land, Appendix A of the advice note in *Design manuals for roads and bridges*
Highways Agency, London

INSTITUTE OF PETROLEUM (1998)
Guidelines for investigation and remediation of petroleum retail sites
Institute of Petroleum, London

A2 Summary of the legislative regime in the UK and other countries

A2.1 THE UK REGIME

A2.1.1 Introduction

Contaminated land in the UK is regulated under several regimes, of which the main ones are:

- Planning
- Part IIA of the Environmental Protection Act 1990 (the contaminated land regime)
- The Integrated Pollution Control regime which is being replaced by the Pollution Prevention and Control regime
- Waste Management Licensing
- The Water Resources Act 1991.

A2.1.2 Planning

Contamination is a material consideration under the planning regime. This means that planning authorities must take the potential or actual presence of contamination into account when considering a planning application for a site that is suspected or known to be contaminated. In such a case, the planning authority may be required to consult authorities such as the Environment Agency. The planning authority may require a developer to investigate, assess or remediate the contamination, for example, as a condition to being granted planning permission. Planning Policy Guidance 23 (PPG23), which is being updated, describes the interaction between planning and contaminated land controls.

The standard of remediation is suitability for the proposed use. This is a risk-based approach that considers the risks posed by contamination on the site to various receptors including but not limited to people, buildings and the environment.

A2.1.3 Part IIA of the Environmental Protection Act 1990 (contaminated land regime)

The purpose of the contaminated land regime, which was implemented in England in 2000, is to remediate land contaminated by historic pollution incidents and which poses an unacceptable risk to people or the environment. Under the regime, local authorities must inspect their areas to identify contaminated land.

Part IIA defines "contaminated land" as:

any land which appears to the local authority in whose area it is situated to be in such a condition, by reason of substances in, on or under the land, that

- *significant harm is being caused or there is a significant possibility of such harm being caused, or*
- *pollution of controlled waters is being, or is likely to be, caused.*

The term "controlled waters" includes groundwater as well as surface waters and coastal waters.

The Part IIA regime is therefore risk-based, requiring the presence of all three components of a pollutant linkage – contaminant, pathway and receptor – before any land can formally be identified as "contaminated land". Furthermore, the pollutant linkage must be "significant", that is, affecting human health, a protected ecological location, various defined categories of property including buildings, crops and livestock, or controlled waters. Where land is identified that meets this test, the local authority (or, in the case of a subset of contaminated land designated as Special Sites, the Environment Agency or the Scottish Environment Protection Agency) will require one or more "appropriate persons" to remediate the land. The remediation standard under Part IIA is the suitability for use of the land – in the sense that the land would no longer fall within the statutory definition of contaminated land. This is assessed in the context of the current use of the land, or any other use which is both likely and which would not require any new or amended planning permission. (This approach contrasts with that under the planning regime, where the focus is on the proposed new use, for which planning permission is being sought).

An "appropriate person" who must remediate contaminated land is one who caused or knowingly permitted the presence of the contaminant that is causing the land to be regarded as contaminated land. If, after a reasonable inquiry, such a person cannot be found, the owner or occupier of the land becomes the appropriate person but only in respect of the land they own or occupy and not for water pollution issues.

A2.1.4 The Integrated Pollution Control regime (to be replaced by the Pollution Prevention and Control regime)

The Environment Agency and the Scottish Environment Protection Agency have the power under Section 27 of the Environmental Protection Act 1990 to take action to remedy harm that has been caused due to the breach of an authorisation which has been granted under the Integrated Pollution Control regime (IPC).

Parliament has yet to transpose the EC Directive on Pollution Prevention and Control into UK law. The legislation is being phased in by industry sector, ending in 2007. Under both IPC and the Pollution Prevention and Control regime (PPC), the Environment Agency may serve an enforcement notice on a person who breaches the terms and conditions of an authorisation (or permit under PPC) or who operates a designated polluting process (or, under PPC, an installation such as a manufacturing facility) without an authorisation or permit.

In addition, persons applying for a permit under PPC will be required to assess the land on which the authorised installation is located, depending on the category of installation, and prepare a report on the condition of the land. When the installation ceases operation, the land must be reassessed and any deterioration in quality must be rectified so that the site of operation is returned to "a satisfactory state".

A2.1.5 Waste management licensing

An operator of a site that is covered by a waste management licence is not permitted to surrender the licence until they have ensured that the site does not pose harm to people or pollution of the environment.

The main enforcement route for dealing with adverse environmental consequences from a licensed waste management activity is through the imposition of conditions on the site licence. If the holder of a waste management licence breaches the terms and conditions of the licence and the breach results in an unlawful disposal of waste, the operator may be required to remove the waste. If the operator causes contamination other than by a breach of the terms and conditions of the licence or other than in respect of any action authorised by the terms and conditions of the licence, they may be required to remediate the contamination under Part IIA of the Environmental Protection Act 1990.

If a person deposits controlled waste, that is, industrial, commercial or household waste, on a site for which a waste management licence has not been issued, they may be required to eliminate or mitigate the consequences of the deposit under Section 59 of the Environmental Protection Act 1990.

The legislation does not provide a standard to which the remediation must be conducted.

A2.1.6 Water Resources Act 1991

If a person causes or knowingly permits a pollutant to enter controlled waters, the Environment Agency may serve a works notice requiring them to remove the pollutant, remedy or mitigate any harm caused by it and, when reasonably practicable, to restore the aquatic environment and any fauna and flora that are dependant on it (Section 161A).

Remediation of polluted water may involve remediation of contaminated land when the controlled waters involved are groundwater and when soil must, for example, be excavated or treated to remove pollutants that are migrating to the water table.

The legislation does not provide a standard to which the remediation must be conducted.

A2.2 THE EUROPEAN UNION

Since 1972, the protection of the environment has been a stated objective of the European Union (EU). Since that time a large amount of legislation has been developed. Most EU environment-related legislation deals with issues surrounding air quality, waste and water. Until now, contaminated land has not been extensively tackled at EU level. However, on 9 February 2000, the European Commission submitted a White Paper that sets out the matters for which various parties might be liable in relation to remedying environmental damage. The paper proposes that the person who controlled the activity that led to the damage should be strictly liable (that is, no fault on the part of the person would have to be proved) for traditional damage (ie personal injury and property damage) and biodiversity damage (damage to natural resources) for a closed list of EU environmental legislation. In addition, fault-based liability would apply in the case of actions outside the closed list that resulted in biodiversity damage. Another party who claimed against the person would need to demonstrate a plausible link between the responsible person's activity and the damage.

It is proposed that the new regime will not impose retrospective liability and will apply to future pollution only. Therefore member states of the EU will continue to address historic pollution by the application of national legislative regimes.

A2.3 OTHER COUNTRIES

Contaminated land has become a high-profile issue for many countries. Strategies for dealing with historically contaminated sites have developed on a national or regional basis. Awareness of the problems caused by contamination is especially high in industrialised countries, where attention is often focused on land contamination either in relation to land regeneration or in response to specific incidents. The approach of other countries to land contamination usually differs from UK policy, legislative regime and general practical approach. Further reading is recommended if the reader needs to gain a complete picture of the applicable regulations.

The problems posed by contamination vary between countries, and different approaches and procedures for assessing and managing land contamination have developed. This resulted both from different political approaches and from responses to specific incidents. For example, in the USA, the Love Canal incident in the late 1970s, when hazardous waste seeped into homes near a former waste dump, triggered introduction of the Comprehensive Environmental Response, Compensation and Liability Act (Superfund Act). In the Netherlands, the legislative regime initially grew from efforts to protect soil in relation to groundwater as a source of drinking water. Later, legislation was influenced by problems relating to housing built on hazardous waste.

Dealing with the pollution of ground and surface water resulting from contamination recurs in most countries, especially where water resources are scarce in some or all areas. Other issues concern facilitating development, protection of residential development. Risks to human health and the environment form the basis of almost all regimes.

The definition of "significant" contamination varies between countries, as does the interpretation of "contaminated site". The use of soil quality values (generic assessment criteria), such as the "Dutch" list, is common. A risk-based approach to land management is generally applied, and this requires remedial action only where contamination poses an unacceptable risk (actual or potential) to human health or the environment. This tends to relate to the intended current or end use of the land. The Netherlands long ago adopted a strategy that enforced stringent cleanup of land for "multi-functionality", although it has backed away from this approach. This strategy is not common elsewhere, where regimes – in France, for example – require remediation by "control of contaminants".

A2.4 FURTHER READING

FERGUSON, C, DARMENDRAIL, D, FREIER, K, JENSEN, B K, JENSEN, J, KASAMAS, H, URZELAI, A and VEGTER, J (eds) (1998)
Risk Assessment for Contaminated Sites in Europe, Volume 1: Scientific Basis
LQM Press, Nottingham, UK

FERGUSON, C and KASAMAS, H (eds) (1999)
Risk Assessment for Contaminated Sites in Europe, Volume 2: Policy Frameworks
LQM Press, Nottingham, UK

FERGUSON, C (1999)
"Assessing Risks from Contaminated Sites: Policy and Practice in 16 European Countries"
Land Contamination & Reclamation, 7 (2), pp 87–108

ULRICI, W (1995)
International experience in remediation of contaminated sites – synopsis, evaluation and assessment of the applicability of methods and concepts
Federal Ministry of Education, Science Research and Technology, Berlin, Germany

A3 Contaminants requiring specialist advice

A3.1 INTRODUCTION

In addition to the general guidance that is available for contaminated land assessment and management, there are several specialist documents applying to particular types of contamination. This appendix reviews the specialist advice that is available for specific topics. Certain types of contaminated site have unique characteristics, due to the nature of the contamination, its mobility or the severity of the hazards posed. To reflect this, the statutory regime to control such sites may vary and guidance has been published to reflect special provisions or precautions required for these sites. Outlined below is some of the guidance available for the principal categories of contamination where unique problems are anticipated and for which individual procedures for managing risks are required. This section provides an overview of these documents; these should be referred to for further detail concerning the specialist issues for individual sites.

A3.2 LAND CONTAMINATED BY BIOLOGICAL HAZARDS

A3.2.1 Where biological hazards are encountered

Most potentially contaminated land is considered in terms of its chemical or physio-chemical (such as asbestos) contamination. However certain types of site should be considered for the hazards posed by biological contamination. In many cases the biological hazard will be associated with micro-organisms, pathogenic bacteria or viruses. Frequently the risks posed are acute, and generally treated as immediate risks, rather than the long-term (chronic) risks normally considered during assessments of human health hazards. They are thus of particular significance during site investigation planning stages.

Micro-organisms are universally present in the soil and under most conditions do not affect humans at all. Most pathogenic organisms are passed directly from human to human, although the dispersal and survival of these pathogens depends greatly on environmental conditions.

Contaminated land assessors would normally be interested in sites where there is a reasonable chance that infectious agents remain alive (or at least dormant with the potential to be activated, as in the case of bacteria spores). The agents may be present in several media as follows:

- soil – where the causal agents of disease only incidentally affect humans, such as tetanus
- animals – where the agents are only occasionally transferred to humans, such as anthrax from cows and sheep
- humans – where the agent may be spread due to direct or droplet contact.

Most human pathogens are short-lived outside the host, although spores are produced by some bacteria as a means of dissemination and survival. These include:

- the aerobic bacilli (eg the agents of anthrax)
- the anaerobic Clostridia (eg agents of gangrene and tetanus).

Spores have no metabolic activity in themselves but act as survival agents in adverse conditions and growth precursor cells when favourable conditions return (Brady, 1974). It has been recorded that spores can yield viable bacteria from soil over 300 years old.

The contaminated land assessor may need to consider microbiological hazards at the following types of site:

- cemeteries and burial grounds
- animal burial sites
- animal product processing works
- docks, warehouses and places where animal products may have been handled
- buildings where animal products have been used in construction.

A3.2.2 Burial grounds

Where disturbance of burial grounds is expected consideration must be given to both aspects of human health and the legislative requirements. With regard to the latter, disturbance of burial grounds is controlled by the Burials Act 1857, which requires the assessor to obtain special permission from Home Office prior to any invasive works. For consecrated ground ecclesiastical law may apply and a Bishops Faculty may be required. When significant disturbance of human remains is anticipated, such as would be required by major redevelopment of a burial ground, it is not advised that human remains be handled by non-specialist personnel. Normally a specialist sub-contractor is employed to remove human remains. The risk assessor should consider whether removal of remains would cause any immediate consequences not only for development staff but also for members of the public living or working in the vicinity.

Within burial grounds, the most common microbial hazards are bacteria, since viruses generally do not to survive in the ground. However, virus material (such as smallpox) may persist in the ground when flesh has not completely decomposed; the assessor should allow for unexpected preservation of human remains. It is usual to consult with the local environmental health officers before conducting invasive investigations of burial grounds.

A3.2.3 Animal product handling, animal burial places

Diseased animals were frequently buried on farmland, either officially (recorded) or unofficially. Animal products arise from the processing of animals at slaughterhouses, rendering, gelatine, glue and meat processing plants. Leaking tanks, pipework or storage chambers and cracks or joints in floor slabs and surfaces may have allowed entry of liquid containing pathogens into the ground. The risk of encountering pathogens on such site diminishes with time because conditions in the soil are usually hostile. Some can persist for several years, however. Anthrax is commonly considered in relation to buried animal carcasses or from areas where animal products have been handled. The routes of infection are via skin contacts (cuts and abrasions), ingestion and inhalation. Other diseases may persist where unfit meat is present.

A3.2.4 Biological hazards in construction materials

Animal products were sometimes used in construction products, particularly in buildings constructed up to the early part of the 20th century. Blood products were used in mortars and plasters, and horsehair was used in plasters. Anthrax spores have been found to be associated with horsehair plaster.

A3.2.5 Available guidance

Department of the Environment (1995)
Industry Profiles, Animals and Animal Products Processing Works
This document provides guidance on the causes of ground contamination due to animal products processing and likely persistence of pathogens.

Further reading

BRADY, N C (1974)
The Nature and Properties of Soils, 8th edition
Macmillan, USA

DEPARTMENT OF THE ENVIRONMENT (1995)
Industry Profiles, Animals and Animal Products Processing Works
Department of the Environment, London

TURNBALL, P (1996)
"Guidance on environments known to be or suspected of being contaminated with anthrax spores"
Land Contamination & Reclamation, vol 4, no 1, pp 37–45

A3.3 SOIL GASES

A3.3.1 Where soil gases are encountered

Gases are normally present within any ground within the interstitial spaces of the soil matrix. They are derived from one or more of the following sources.

Atmospheric gases. Gases from the earth's atmosphere enter the ground through surface cracks, pores and fissures and become mixed with other gases in the soil matrix.

Gases from natural strata. Gases may be present in the ground as a result of their release from natural strata, including peat, coal measures, crude oil and natural gas strata, estuarine, river, lake or dock sediments and other naturally enriched organic strata. Release of these may be an ongoing natural process. However this can often be enhanced by man's activities, such as mining or groundwater pumping, which release pressure on gas-containing rock, creating conditions for the liberation of gas into the artificial void.

Gases from manufacturing or storage processes. Gas may be released into the ground as a result of the deposition of materials (land-filling or land-raising), leakage of raw or manufactured materials from storage vessels and the generation of gases as a result of chemical (or biological) reaction of buried materials with the ground.

The composition of soil gases varies due to many factors, such as variation in atmospheric conditions, changes in groundwater level and degree of biological activity and source. Primarily attention is given (particularly in the context of landfill sites) to the gases methane, carbon dioxide and oxygen.

A3.3.2 Pathways for gas migration

Soils gases are often highly mobile and may migrate through the ground, dependent on the relative porosity of the ground and availability of preferential pathways. The movement of gases through the ground is governed either by diffusion or pressure-driven flow. Gas pressure will only build up if generation rate exceeds diffusion rate. Gas may become dissolved (under pressure) in solution (such as in groundwater); in these situations the gradient of groundwater will control gas migration. Gases may enter buildings and structures for example through a poor quality (cracked or honeycombed) concrete floor slab or via services penetrating the slab.

A3.3.3 Receptors

The soil gases considered in this report are of a generally potentially hazardous nature. Particular hazards to humans arise from the gases' toxicity, flammability/explosive and asphyxiant properties; to structures (including buildings and other development), their flammability and corrosive properties; and to the environment, their toxic effects and retardation of plant growth. In many cases the risk is acute, such as the potential for methane to ignite in confined spaces. The available guidance reflects these risks.

A3.3.4 Available guidance

Measurements of soil gases are normally made where these are expected on a particular site, established usually by preliminary researches into the land-use history of the site and its environs. This is necessary to assess the risk from soil gases to specific activities at the site. In most cases, the scenarios relate gas being present due to waste disposal activities. Several documents, below, set out the significance of the gas measurements and recommended action. Key legislative control is not by regulations specific to soil gases, but by general contaminated land and also waste management legislation.

Department of the Environment Waste Management Paper No 26A: Landfill Completion

Sets out the technical issues to be considered for the surrender of a landfill waste licence, and details the criteria for completion in relation to measurements of gas and leachate. Particular criteria are included for comparison with the collected data:

- measurements of gas concentration should not exceed stated methane and carbon dioxide concentrations over a 24-month period subject to stated atmospheric conditions
- Measurements of gas emission rate should be made for comparison against stated criteria
- Interpretation of data should take into account natural conditions, especially where these are likely to be greater than the stated completion target criteria.

Department of the Environment Waste Management Paper 26D: Landfill Monitoring

Details monitoring requirements for landfill sites and gas measurement techniques, which will be relevant for the risk assessment process at the data collection phase.

Department of the Environment Waste Management Paper 27: Landfill Gas

Gives guidance on the requirements for the monitoring of landfill sites including the parameters to be monitored, the location and spacing of monitoring points around landfill sites and where landfill gas is expected due to migration. The document states limits in relation to significant concentrations of soil gases in buildings, at 1 per cent by volume methane and 1.5 per cent by volume carbon dioxide. The guidance gives recommendations for development within a 250 m radius of a landfill.

Methane and Other Gases from Disused Coal Mines. The Planning Response Technical Report and Summary Report (DETR 1996)

This report describes the hazards to new development associated with the emission of gases from mines. The origin of mine gases is reviewed and the pathways for these gases to migrate to the surface. Guidance is given as to the identification of emission hazards and assessment of risk. Control measures are recommended for incorporation into building design, within the planning framework.

Building Regulations Approved Document C2 Section 2 (1992)

This guidance outlines the potential risks to buildings from landfill gas and the general approach to engineering mitigation to be considered at building design stage.

CIRIA Report 130: Methane: its occurrence and hazards in construction (CIRIA 1993)

Details the hazardous properties of the methane and other gases originating from landfill or natural sources and provides guidance to enable situations to be identified where methane may arise where this will have consequences for construction.

CIRIA Report 131: The measurement of methane and other gases from the ground (CIRIA 1993)

Provides guidance on the collection of soil gas monitoring data. Techniques for measuring soil gases within the ground and within buildings are described.

CIRIA Report 149: Methane and associated hazards to construction: protecting development from methane (CIRIA 1995)

The report considers the hazards to buildings that arise from the presence of landfill gases within the soil and sets out the principles for the design of protection measures that should be incorporated into potentially affected buildings.

CIRIA Report 151: Methane and associated hazards to construction: interpreting measurements of gas in the ground (CIRIA 1995)

Guidance is provided to enable a site assessor to interpret data collected within investigations instigated, eg, in accordance with CIRIA Report 131. The report describes the process of interpreting gas measurements and the likely validity of the results.

CIRIA Report 152: Methane and associated hazards to construction: risk assessment for methane and other gases from the ground (CIRIA 1995)

Within this report there is detailed a methodology for the assessment hazard and risk associated with soil gas regimes. Some overseas experience is described. The report relates the communication of gas-derived risk to risk perception.

BRE Report 212: Construction of New Buildings on Gas-Contaminated Land (Building Research Establishment, 1991)

Considers soil gas present in ground from both natural and man-made sources. The report discusses the background on the production of landfill gas, its properties and potential to migrate off site. The assessment of soil gas contamination is discussed. The document refers to the "trigger" values given in the Approved Document C (see above) and Waste Management Paper 27 (that is 1 per cent and 1.5 per cent by volume of methane and carbon dioxide respectively). The options for mitigation at design stage for different types of new development are outlined.

BRE Information Paper: Fire and Explosion Hazards Associated with Redevelopment of contaminated Land (Building Research Establishment, 1987).

The paper reviews the explosion and toxic hazards that may be associated with combustible material, such as colliery spoil and domestic refuse. Fires at contaminated sites are discussed, including ignition sources, methods by which the combustibility of material may be assessed and methods for fire prevention.

A3.4 RADIOACTIVE HAZARDS

A3.4.1 Where radioactive materials are encountered

All sites have background radiation levels contributed to from world-wide nuclear weapons testing and discharges from nuclear facilities. There is hence a natural background level within the UK against which radiological risks are assessed.

A3.4.2 Sources

Elevated radioactive contamination may be present on a site due to one or more of the following scenarios:

- as a result of airborne or water-borne deposition of materials from (nearby):
    - nuclear reactors
    - nuclear research facilities
    - sites where nuclear materials have been stored
- as a result of airborne deposition of materials from machining of depleted uranium (used in some conventional munitions)
- diffuse ground contamination in the areas that non-nuclear industry radioactive materials were manufactured, processed, stored or used, eg fluorescent paint and some specialist metallic alloys
- point source contamination from misplaced or buried radioactive sources from x-ray sets or laboratories
- contamination of the fabric of buildings by radioactive materials, due to accretions of dusts or spillages, in buildings used to manufacture or process radioactive materials
- naturally derived radioactivity, such as radon gas. Radon is part of the radioactive decay chain of Uranium-238 and occurs in parts of England and Wales. Natural exposure is sometimes sufficiently high to present a risk.

In many cases, radioactive materials are significantly toxic. For substances with low radioactivity, toxicity may be the primary risk.

A3.4.3 Pathways

The pathways for the risk presented by radiological materials are identical to those for other soil contaminants, ie ingestion of soil or plants grown therein, dermal contact or inhalation of dusts or vapours as discussed in Chapter 5. However, the relative weighting of risk attributable to each pathway may be different, with increased risk of harm from inhalation of dusts and dermal contact.

A3.4.4 Receptors

The risk to receptors is both radiological and toxic. The potential receptors are humans, fauna and flora. At the levels likely to be found as a result of gradual contamination, the principal health risks are chronic, with increased cancer incidence being the main radiological risk.

A3.4.5 Available guidance

Risk assessment for radiological contamination is a highly specialised and it is advised that specialist advice be sought.

Under UK law relating to nuclear activities, radiological contamination caused by or due to the nuclear industry (civil and military) is firmly the responsibility of the polluter to remediate. Thus for the nuclear industry contamination scenarios, risk assessment will generally be performed by or on behalf of a nuclear industry operator. The operators have their own (generally not publicly available) guidance documents for the performance of risk assessments. Regulations for the management of radioactively contaminated land have been drafted by DETR (1998).

Further reading

Guidance documents relevant to risk assessment of radiologically contaminated land include:

DETR (1998)
Control and Remediation of Radioactively Contaminated Land – A Consultation Paper
DETR, London

ENVIRONMENT AGENCY (2000)
Technical Support Materials for the Regulation of Radioactively Contaminated Land,
R&D Technical Report P307 (P3-055)
Prepared by Entec UK Ltd and National Radiological Protection Board.
Environment Agency, Bristol

The Ionising Radiations Regulations 1985
HMSO, London

NRPB (1998)
Radiological Protection Objectives for Land Contaminated with Radionucleotides
NRPB, Didcot, UK

TILL, J E and MOORE, R E (1988)
"A pathway analysis approach for determining acceptable levels of radionucleotides in soil"
Health Physics, 55 (3) 1988, pp 541–548

Much guidance has been published about risks due to radon. The papers below describe the sources of radon, the associated risks and means by which it is measured.

BRE (1992)
Radon Sumps: a BRE Guide to radon remedial measures in existing buildings,
Guide BR227
Building Research Establishment, Watford, UK

NRPB (1993)
Radon Affected Areas: Scotland, Northern Ireland, NRPB 4, No 6
National Radiological Protection Board, Didcot, UK

NRPB (1996)
Radon Affected Areas: England, Wales, NRPB 7, No 2
National Radiological Protection Board, Didcot, UK

SCIVYER, C R and GREGORY, T J (1995)
Radon in the Workplace: a guide to radon measurement and remedies for non-domestic buildings, Guide BR293
Building Research Establishment, Watford, UK

USEPA (1993)
Protocols for Radon and Radon Decay: Product Measurements in Homes,
EPA 402-R-92-003
United States Environmental Protection Agency, Office of Air and Radiation

USEPA (1993)
Indoor Radon and Radon Decay: Product Measurement Device Protocols,
EPA 402-R-92-004
United States Environmental Protection Agency, Office of Air and Radiation

WELSH, P, PYE, P W and SCIVYER, C R (1994)
Protecting Buildings with Suspended Timber Floors: a BRE Guide to radon remedial measures in existing dwellings, Guide BR270
Building Research Establishment, Watford, UK

A3.5 MUNITIONS AND EXPLOSIVES

A3.5.1 The importance of munitions and explosives assessment

Munitions and explosives present an acute risk within contaminated land management and present a particular problem to the construction industry. Encounters with explosives and munitions are associated with risk of fire, explosion and toxicity.

A3.5.2 Sources

Munitions or explosives may be present on a site due to one or more of the following:

- aerial bombing, mainly during the Second World War. Some 10–15% of devices dropped on the UK failed to explode, and many still lie buried where they fell
- deliberate or negligent disposal by burial of munitions/munitions components or explosives in or near former defence sites
- accidental loss and surface burial of munitions/munitions components or explosives in the areas they were manufactured, processed, stored or used

- diffuse ground contamination presenting primarily a toxic risk, in the areas explosives were manufactured, processed, stored or used
- contamination of the fabric of buildings by explosives, due to accretions of dusts or spillages, in buildings used to manufacture or process explosives.

The potential for the presence of the above risk sources is significant at or near present or former defence sites. These include:

- sites owned at any time by the ministry of defence or its predecessors, including former military airfields, dockyards and barracks
- present or former munitions or explosives production factory sites
- present or former munitions/armaments storage areas, testing areas or ranges
- sites subjected to aerial bombing during the Second World War.

Munitions/munitions components vary in size from 5 mm to 1.5 m. Explosives are divided into inorganic and organic materials. Inorganic explosives are generally not environmentally persistent, but break down into non-energetic compounds (which may still be toxic due to heavy metals). Organic explosives make up the bulk of explosives used since 1860, and are generally highly persistent in the environment. Some are significantly toxic. Discrete particles or fragments of explosive may in limited circumstances present a source of explosion or fire. In some locations, there is the possibility of munitions being filled with chemical weapons agents.

A3.5.3 Pathways

Munitions and discrete explosive fragments are generally immobile as environmental contaminants, with a pathway present only by active disturbance by a receptor.

The pathways for the toxic risk presented by explosive materials are identical to those for other soil contaminants, ie ingestion of soil or plants grown therein, dermal contact or inhalation of dusts or vapours.

A3.5.4 Receptors

The risk to receptors from munitions and discrete explosive fragments is acute, with damage to humans, animals, plant or buildings occurring when disturbance leads to an explosion or deflagration. Even the smallest devices are capable of causing serious injury, and although munitions often become less subject to accidental detonation with age, the damage potential usually does not reduce. The toxic risk is to human health. Little information is available assessing toxicity to flora, although a basic assessment of their chemistry indicated that most organic explosives are unlikely to be significantly phytotoxic.

A3.5.5 Available guidance

Very little publicly available guidance exists covering munitions as an environmental contaminant, perhaps because risk assessment where the primary risk is human injury from the explosion of a military device is considered a specialist activity. Certain guidance is implicit within legislation, as below.

The Construction (Design & Management) Regulations 1994, which apply to any intrusive site investigation works performed as part of a risk assessment (as well as to any subsequent remedial or construction works). These require that a health and safety plan is prepared covering site works, approved by a planning supervisor who is competent to assess the risks to the proposed works. This in effect requires any site-based element of a contaminated land risk assessment to involve a planning supervisor experienced in munitions risk assessment.

The Packaging of Explosives for Carriage Regulations 1991 and Road Traffic (Carriage of Explosives) Regulations 1989 effectively prevent the off-site transport by civilians of any munitions discovered during an on-site risk assessment. Exemptions from these regulations can sometimes be obtained by specialist companies from the Health and Safety Executive. For low numbers of finds, the police or armed forces will usually respond by removing or destroying suspected live munitions found during site works.

For the site investigation phase of a risk assessment for the presence of munitions, a major part is usually played by non-intrusive explosive ordnance detection (EOD) techniques. There is a broad body of publications detailing methodologies for performing such investigations (and the subsequent disposal of munitions discovered). These publications are not publicly available, and are of limited use other than to the military or specialist EOD companies, since they describe procedures involving risk of explosion.

Guidance covering both munitions and toxic explosive issues is in preparation on behalf of the Environment Agency for its own use, and it is not clear whether this will be made publicly available. The documents are:

- *Land Contamination: Technical Guidance on Special Sites:*
    - *R&D Technical Report P152 – MoD Sites*
    - *R&D Technical Report P153 – Chemical Weapons*
    - *R&D Technical Report P154 – Biological Weapons*
    - *R&D Technical Report P155 – Explosives Manufacturing and Processing Sites.*

The documents give guidance for EA staff acting in an advisory or regulatory role in respect of the provisions of the Special Site Regulations within Part IIA of the Environmental Protection Act, which came into force on 1 April 2000. The documents describe the characteristics of the types of site, the type and nature of likely contamination, and the issues relevant to successful characterisation, evaluation and remediation.

Where the principal risk is toxicity, rather than any radioactive hazard, the procedures reviewed in Chapters 5 and 6 are applicable. Additional documents of interest are listed below.

DETR (1995)
Industry Profile – Chemical Works: explosives, propellants and pyrotechnics manufacturing works
Building Research Establishment, Watford, UK.

DETR (1995)
DoE Industry Profile – Engineering Works: mechanical engineering and ordnance works
Building Research Establishment, Watford, UK.

These documents are part of a series describing the contaminative risks associated with particular industries. They describe the types of sites, the manufacturing processes and materials used. The publications list chemicals (including explosives) that may be present as contaminants, and comment on contaminative pathways.

ROYAL ORDNANCE (1990)
Explosive Materials – Determination of Toxicological Hazards and other Properties
Royal Ordnance, UK

This publication generates land and groundwater contamination risk assessment threshold values for 11 common organic explosives. The document is somewhat dated, but is still the only UK source of trigger concentrations for explosives. The values are conservative, and are generally accepted by UK regulatory authorities.

A3.6 ASBESTOS

A3.6.1 The importance of asbestos contamination

Asbestos contamination can be found on many sites and in many forms. Asbestos occurs as fibrous crystals that are capable of being woven into fabric, and its fire retardant and insulation properties are well known. Asbestos has also been combined into building materials and paints, in particular asbestos cement board.

Asbestos presents a hazard to human health via the respiratory system. Inhalation of asbestos fibres is related to several respiratory diseases including asbestosis, lung cancer, mesothelioma and pleural disease. Asbestos persists indefinitely as an environmental contaminant.

A3.6.2 Sources

Asbestos may be present on a site due to one or more of the following:

- burial of asbestos lagging
- burial or surface scatter of products containing asbestos (particularly cement-bonded asbestos products used as building materials)
- diffuse contamination ground resulting from surface deposition of fibres from damaged fibrous asbestos lagging or from former asbestos manufacturing sites.

The potential for the presence of the above sources is significant in nearly every former industrial site, and in some older residential properties. The materials were intensively used in the UK for more than 100 years as thermal insulation of steam pipes and buildings, and in cement-bonded form or as additives in (primarily) building materials. Of particular risk are former government-owned sites (especially MoD sites), which used asbestos materials extensively and often continued on-site burial practices for disposal of asbestos materials long after legislation prevented this in the private sector.

The main forms of asbestos used in the UK (in descending order of risk) were:

- crocidolite (blue asbestos)
- amosite (brown asbestos)
- chrysotile (white asbestos).

A3.6.3 Pathways

The nature of the risk from asbestos is by the inhalation of fibres only. There is therefore no requirement to assess other exposure pathways. Inhalation can be direct (from asbestos products or contaminated soil) or indirect (from clothing contaminated by contact with asbestos or asbestos-contaminated soil).

A pathway has been demonstrated where low levels of asbestos as a soil contaminant can (in dry conditions) be a source of significant levels of airborne asbestos fibres.

Despite its nature as a fibrous solid, asbestos is mobile as a ground contaminant. Surface contamination of open ground by deposited lagging can result in significantly elevated levels asbestos fibres up to 1 m below the surface after several years.

A3.6.4 Receptors

The risk from asbestos is to humans only, by long term (10 years or more) exposure to airborne asbestos fibres. The damage caused is irreversible and often fatal. There is evidence to suggest that (repeated) exposure even at very low levels carries some risk of causing serious medical conditions.

A3.6.5 Available guidance

As a bulk waste product and engineering material, a considerable body of specific legislation and documents also exist in reference to risk assessment of occupational exposure to airborne asbestos fibres. This is not of direct relevance to risk assessment of contaminated land, but would apply to remedial works, or possibly if asbestos is encountered during intrusive site investigation.

There is no recent guidance on acceptable levels of asbestos in soil in the UK. The situation is further complicated by difficulties in quantitative analysis of asbestos in soil. The test in common use for asbestos as a soil contaminant in the UK has a reporting threshold of 0.1%. However, most laboratory analysis does not achieve the 0.1% threshold; 1% is commonly reported. This 0.1% value is the level at which material is defined as "special waste" within UK legislation, that is, the level at which a special hazard exists. Hence, lower levels that may still be hazardous are often ignored by risk assessments currently performed.

The following documents are considered of direct relevance to risk assessment of contaminated land where a risk from asbestos is suspected.

DETR (1995)
Industry Profile – Asbestos manufacturing works
DETR, UK

One of a series of publications describing the contaminative risks associated with particular industries. It describes the history of the asbestos manufacturing industry in the UK, detailing the manufacturing processes and materials used. The document gives useful information on the range of products that contained asbestos, although it is primarily focused on the direct risk from former asbestos-manufacturing sites.

DEPARTMENT OF THE ENVIRONMENT (1990)
Inter-Departmental Committee on the Redevelopment of Contaminated Land Guidance Note 64/85– Asbestos on contaminated sites, 1990

Comprehensively discusses the process of risk assessment of contaminated land for asbestos. It describes the form of materials, types of sites, and quantitative risk assessment and advises that concentrations of asbestos in soil of 0.001% should be considered to present a significant risk. This level is only one order of magnitude above the naturally occurring level of asbestos in soil, and is beyond the detection threshold of any routine test available.

DEPARTMENT OF THE ENVIRONMENT (1979)
Waste Management Paper No 18, Asbestos Wastes – A technical memorandum on arisings and disposal including a code of practice
HMSO, London

This document gives guidance for the safe disposal of asbestos-bearing wastes.

Table A3.1 indicates the primary purpose of the published guidance for special topics.

Table A3.1 *Primary purpose of published guidance for various special topics*

Document	Site investigation planning	Monitoring and evaluation	Hazard assessment	Remedial design
DoE WMP26A				
DoE WMP26D	✔	✔		
DoE WMP27	✔	✔	✔	
Methane and Other Gases from Disused Coal Mines	✔	✔	✔	✔
Building Regulations Approved. Document C	✔	✔	✔	
CIRIA Report R130	✔	✔	✔	
CIRIA Report R131	✔	✔		
CIRIA Report R149	✔	✔	✔	✔
CIRIA Report R151	✔	✔	✔	
CIRIA Report R152			✔	✔
BRE Report 212		✔	✔	
BRE Information Paper Fire and Explosion Hazards			✔	

Table A3.1 *Primary purpose of published guidance for various special topics (cont)*

Document	Site investigation planning	Monitoring and evaluation	Hazard assessment	Remedial design
Documents on asbestos				
DoE Industry Profiles	✔			
ICRCL Asbestos on Contaminated Sites	✔	✔	✔	✔
Munitions and explosives				
Land Contamination: Technical Guidance on Special Sites	✔	✔	✔	✔
Royal Ordnance – Explosive Materials – Determination of Toxicological Hazards and other Properties, 1990	✔	✔	✔	
Radioactivity				
NRPB Radiological Protection Objectives for Land Contaminated with Radionucleotides.			✔	
Land Contamination: Technical Guidance on Special Sites:	✔	✔	✔	✔
R&D Technical Report P158 – Nuclear Establishments				

A4 Ecological risk assessment

A4.1 WHAT ARE ECOLOGICAL RISK ASSESSMENTS?

An ecological risk assessment (ERA) will predict actual or potential harm caused to flora, fauna in selected ecosystems based on the known chemical and toxicological characteristics of the contaminants of potential concern. ERAs are undertaken in a slightly different way to human health risk assessments, being concerned with the effects of contamination on more than one species. ERAs therefore often consider the inter-relationships between species in what can be very diverse communities and ecosystems. The ecotoxicity of a chemical is seen to vary both inter- and infra-specifically such that animal and plant species are differentially sensitive to individual chemicals. Furthermore, individuals within a given population (of a single species) may also be differentially sensitive according to their maturity, sex, size and baseline health status etc.

The term "ecosystem" refers to a given community (ie a defined group of populations) within its habitat. Consideration should be given to the fact that only selected ecosystems are regarded as receptors within the Part IIA of the new contaminated land regime, and certain ecosystems may also be protected under other legislation. In some circumstances, the flora or fauna will have other, more general, significance such as rarity, public perception or domestication. Under Part IIA of the Environmental Protection Act 1990 the ecosystems classified as receptors are very specific, and these are listed in Table 2.1.

A4.2 WHY ARE ECOLOGICAL RISK ASSESSMENTS UNDERTAKEN?

Ecological risk assessments are carried out where there is reason to expect that contamination in soil and water may be significantly affecting ecosystems on or off the site. Another consideration is the potential for construction work undertaken during remediation or redevelopment to affect the ecology of the area adversely. Ecosystems are protected under a variety of environmental legislation; regulators, local authorities and pressure groups may also take an active role in protecting certain ecosystems within their region.

ERAs are frequently carried out in the construction/remediation context because disturbance of the ground by construction works can cause contamination to be mobilised and creates new pathways for contaminants to reach receptors. There is, therefore, potential for significant harm to be caused to flora and fauna in varying degrees during construction activities. ERAs are often undertaken as a formal part of the environmental impact assessment process for certain types of development project. They may be used to provide information on which decisions on more general developments can be based.

In terms of construction, examples of likely impacts of contamination on the ecology are shown in Table A4.1

Table A4.1 *Examples of possible impacts of contamination on ecology*

Scenario	Possible impact
The presence of contamination on areas to be landscaped or developed as domestic gardens.	Phytotoxic metals in the ground may prevent healthy plant growth. This may result in the requirement for remediation or removal of contaminated soil to landfill and the importation of clean topsoil, with cost implications on the proposed development.
Construction work causing new contamination sources, such as the spillage of fuels, cement, solvents etc.	Contamination introduced to previously unaffected areas.
Construction work creating new migration pathways, such as: • surface runoff increasing due to the compaction of the ground • foundations and underground services providing preferential migration routes.	Contamination introduced to previously unaffected areas.

In some cases, flora and fauna may present a hazard to other ecosystems or even proposed construction work. For example, Japanese knotweed competes aggressively with other plants for water and nutrients and can cause structural damage to foundation work.

The findings of the risk assessment may highlight unforeseen development costs, so should ideally be undertaken at the pre-feasibility or feasibility stages of design.

A4.3 THE SCOPE OF AN ECOLOGICAL RISK ASSESSMENT

In general terms, the ERA should consider the types of contamination present and the way that individual contaminants will behave in and move through the environment being studied. It should describe the ecotoxicological impacts of any contamination reaching an individual organism, the population and the community at risk. The ERA should also consider inter-relationships between chemical effects and the species at risk, since there will commonly be more than one contaminant present and more than one species of plant or animal impacted. It is often the complexity of characterising these inter-relations (multiple chemicals and multiple species) that makes ERA a specialist and highly interpretative process.

There are few methodologies available for undertaking a formal ERA. Often the type of assessment made depends on the type of contamination present and the significance of the ecosystem thought to be at risk. It is important to set the ecosystem under scrutiny within the context of the local area. For example, the impact of contamination on the ecology of a heavily industrialised inner city area might be considered of lesser significance than that within a recreational park or forest.

A4.4 HOW TO CARRY OUT AN ECOLOGICAL RISK ASSESSMENT

As with other forms of contaminated land risk assessment, ERAs are usually undertaken using a tiered approach where successive levels of sophistication and detail require an increased amount of technical data on which to base the assessments.

Initial survey and conceptual site model

The initial task usually involves consideration of the types of ecosystem potentially at risk. This can be carried out by looking at the general characteristics of the area and undertaking a site survey to determine the variety of ecosystems present, the populations that comprise them and obvious signs of adverse ecological effects. Some form of initial site survey will normally be carried out, by an experienced ecologist, to determine the ecosystems present. It is important to appreciate that adverse ecological effects observed may not necessarily be directly or even indirectly related to contamination. A multitude of environmental, physical and biological factors, including general disturbance by man, may influence the health of an individual organism, a population or community structure. For example, die-back of vegetation on a landfill site may be attributable to the migration of landfill gas into the near-surface soils, but might just as easily be a function of the dry soil conditions in well-draining, uncompacted waste there.

As with other forms of risk assessment, a conceptual model should be developed to permit visualisation of the relative relationships between the contaminants, the pathways and the sensitive ecological receptors. Some specialist knowledge of ecology and ecotoxicology is required so that the complexities of the inter- and intra-specific variability in organism, population and community response can be characterised.

Direct comparison with soil screening levels

Where the initial site survey and conceptual site model indicate that a potential risk to ecological systems exists, data should be collected on contaminant concentrations in soils and water, and on the physical and chemical characteristics of the soil and groundwater on the site. The ecosystems present should be surveyed in detail.

The simplest approach to assessing the risks posed to ecosystems is to compare observed concentrations of contaminant in soils and water on and off the site directly with generic assessment criteria (see Chapter 6). Ecological risk-based guidelines against which environmental concentrations may be compared include environmental quality standards (EQSs) for chemicals of potential concern and other criteria that may be readily available for this purpose (ie risk-based ecotoxicological data derived specifically for use in assessing contaminated soil). More generally, it is possible to use ecotoxicological data from the literature to compare measured concentrations of contaminants with known and predicted toxicity data. It is important to appreciate that in some cases the toxicological impact of chemicals on a certain species may be additive – two or more chemicals may cumulatively affect the health of an organism.

Deriving criteria, use of models

Where it is not possible to record accurately the concentrations of individual contaminants at the ecosystem point of exposure, it may be possible to use some of the more common fate and transport models to estimate environmental concentrations as necessary. Once the concentration of selected contaminants at the point of ecosystem exposure has been estimated, then comparison with EQSs and other ecotoxicological data in the literature can be used to derive predictions on ecosystem impact.

Ecotoxicological assessment criteria are generally derived from toxicity test data collected from laboratory studies undertaken on individual species and where effects on one or more species are extrapolated to predicted effects on populations, communities and on ecosystems. This approach can produce highly site-specific data, but must be undertaken with caution, making allowance for the differences between trial and field conditions. For example there may be difficulty in simulating the behaviour of species in relation to feeding patterns, or predicting the relationship between species. It is important to appreciate that where toxicological data for one species exists it is not always possible to predict the affects of contamination on other species, even where the species of concern are of the same genus or similar.

Ecotoxicological assessment criteria are derived in a similar fashion to human toxicological data. They are based on published toxicity data. Ecotoxicological data is usually published in the form of one of the following:

LC50 (where LC = lethal concentration) defines the concentration in a medium that is lethal to 50% of the given population exposed. These tests are usually further classified by length of time of exposure, ie 24, 48 or more normally 96 hours.

LD50 (where LD = lethal dose) defines the concentration of a single dose that is lethal to 50% of a given population exposed.

EC50 (where EC = effective concentration) defines the concentration required to elicit a given adverse effect (reduction in luminescence or immobility etc) in 50% of the individuals exposed. Again, the exposure period is usually defined

NOEC (where NOEC = no observed effect concentration) defines the concentration at which no adverse effect has been recorded (ie this could be the lowest of all EC50s stated for a given species across a number of effects).

Box A4.1 gives an example of an ecological risk assessment.

Further reading

ENVIRONMENT AGENCY (*in preparation*)
"Assessing risks to ecosystems from land contamination"
Project P5-029

Box A4.1 *Example of an ecological risk assessment*

Assessment of former sawmill site

Background. The site of a former sawmill was proposed for redevelopment. As part of the planning conditions a Phase 1 study with site walkover visit was undertaken.

Industrial units surround the sawmill to the north, east and west. To the south is an undeveloped, low-lying, occasionally marshy area with a stream flowing east to west. The local public perceives this undeveloped land as an area of natural beauty (although it does not hold any such legal status).

Historical studies. The Phase 1 study indicated that a sawmill had existed at the site between 1910 and 1980 Before 1910 the site was undeveloped. Historical maps indicated little change in the layout of the site over the years. The sawmill had been centred on three buildings, and the footprint of these was visible at the time of the site visit. Two tanks were indicated on the maps.

Published geological maps and other records indicated that the geology at the sawmill site comprised alluvium with gravel (to a maximum depth of 5 m) overlying clay. The hydrogeological gradient appears to be towards the marsh in the south.

The site walkover visit. A contaminated land scientist and an ecologist undertook a site walkover. Above-ground fuel tanks were observed at two locations across the site. The tanks were empty. No evidence of any bund or other containment method was observed. The ground below the tanks was severely stained and little vegetation was growing.

A severely cracked concrete pit was observed within the largest building. Dark green staining was observed both within and around the pit. The Phase 1 study determined that this was this was the location of the wood treatment process. Site records indicated the most commonly used wood treatment product used at this site had been tanalith (35% copper sulphate, 45% sodium dichromate and 19% arsenic pentoxide). Also, creosote had been widely used.

Significant plant species (significant more in terms of public perception than rarity) were observed both on the site and at the undeveloped land to the south of the sawmill. These included yellow loosestrife, cross-leaved heath and wild angelica. Several areas of distressed vegetation were also observed. Numerous aquatic invertebrate and vertebrate species were identified, the largest of which was the minnow (Phoxinus phoxinus).

The desk study concluded that several potential contamination sources were observed at the sawmill and that the type of ecology observed at the sawmill and surrounding land was determined to be significant. The limited information from the desk study would appear to indicate the potential for a contamination pathway from the contamination sources (fuel and wood treatment) to potential receptors (the watercourse and flora). An intrusive site investigation, sampling and testing (both in the sawmill site and in adjoining land) confirmed the site geology, hydrogeological regime (that groundwater flowed from the north to the south, towards the stream) and the presence of elevated concentrations of fuel oil and heavy metals (arsenic and copper) within the soil and groundwater. Surface water monitoring was undertaken both upstream and downstream of the sawmill over a six-month period. It found slightly elevated concentrations of fuel oil and heavy metals (arsenic and copper) within the surface water (downstream compared to upstream of the sawmill).

Analysis of the data indicates that contaminants were migrating at varying rates off site towards the stream and the low-lying marshy ground. Groundwater was encountered less than 0.2 m below ground level in this area.

It was concluded that, for ecosystems at and adjacent to the site, contaminant-pathway-receptor linkages were complete and that there was potential for adverse effects.

The effects on plant life were determined using field study techniques. The type, number, general condition and distribution of plant life was examined at two areas:

- Area 1 – located within the marshy undeveloped area but outside the influence of the identified contaminated groundwater plume
- Area 2 – located within the marshy undeveloped area and at the approximate centre of the identified contaminated groundwater plume.

The range and general condition of plants species identified in Area 1 were greater than those identified in Area 2. Distressed vegetation was determined in Area 2.

The potential effects of contamination on the freshwater fauna were determined by comparing observed surface water concentrations to the published LD_{50} concentration for minnows. The mean arsenic concentration determined during the surface water-monitoring program was slightly elevated above the LD_{50} for minnows (for a 100 m stretch of the stream down-gradient of the sawmill). Concentrations of copper and fuels were significantly below the LD_{50} for minnows.

Risk evaluation The assessment concluded that contamination from the sawmill is likely to be impacting upon vegetation and invertebrate life in the adjoining land. As part of the redevelopment strategy, it was recommended that measures were taken to prevent further migration of contamination. During the development the most contaminated soil identified by the investigation was removed for off-site disposal. Post-construction monitoring of groundwater and surface water has shown an ongoing reduction in contamination concentrations.

A5 Estimating risks to building fabric and services

A5.1 WHY ASSESSING RISKS TO THE BUILDING FABRIC AND STRUCTURES IS IMPORTANT

Construction materials are susceptible to attack from aggressive substances within contaminated soils, groundwaters and from vapours and gases (see Appendix 3 on explosive and asphyxiant risks due to soil gases). Concern over attack of materials and structures is chiefly economic, but includes the potential indirect effects on human health, particularly where structures are rendered unstable. The general degradation of materials affects all aspects of life and is well known, particularly in relation to manufacturing and storage. There is no universal solvent, nor any substance that cannot be dissolved into, fused with or otherwise degraded by another substance.

Assessment of risks to construction materials will follow the by now-familiar pattern of identifying the hazard (*contaminant*) that could affect the structure or services, the structure or services at risk (*receptor*), and the linkage between them. In many cases, the contaminant will be in direct contact with the construction materials, such as contaminated backfill placed around foundations, or foundations such as piles placed within aggressive ground. In addition, the structure of the building, and particularly the services, can act as pathways for contamination to migrate. This can occur along trenches within which services are laid and in doing so may be transferred from its original location to a sensitive receptor. The structure or materials are thus effectively the *pathway*.

Sometimes the effects of contamination are obvious. Existing development at the site may exhibit signs of contaminant-generated damage such as cracking, spalling and discoloration. However, expert assistance is needed to determine whether the observed distress is caused by ground contamination or other sources. If the site is to be redeveloped using similar construction materials it may be judged that damage is likely to occur to these materials from the identified ground conditions. Where new development is planned on sites exhibiting no signs of distress to construction materials, risk may be assessed by comparing observed soil concentrations to generic guidance values, if available.

A5.2 WHAT MATERIALS ARE CONCERNED?

A5.2.1 Factors affecting materials attack

This section considers the effects of contaminated ground on some common construction materials. Normally for attack to occur the following conditions have to be satisfied:

- the contaminant must be mobile. Often it will be water-soluble and therefore be transported in the groundwater
- the contaminant must be available in sufficient quantities to have an effect
- there must be contact between the structure/material and the contaminant
- temperature of the ground
- the presence of another chemical such as oxygen either in a catalysis role or else directly contributing to the attack
- materials in the structure are prone to attack or sensitive to the contaminant.

Most construction activities make use of concrete and steel in one form or another. In addition, rubbers and synthetic materials are increasingly used. The following classes of construction material are most commonly encountered on construction sites:

- concrete
- metals
- plastics
- rubbers
- masonry.

Below is a discussion on the various factors relating to durability of these materials.

A5.2.2 Effects on concrete

Concrete is one of the most widely used construction materials used worldwide. When properly manufactured it can exhibit exceptional durability. Basic mortar has survived in structures of Roman origin. The extent to which concrete resists attack from external substances is determined primarily by its quality, which itself is influenced by the design of the mix, accurate proportioning (batching) of the constituents and adequate compaction during placing. Manufacture of concrete under careful control can thus reduce permeability of the concrete in turn increasing is resistance to chemical agents. A wealth of standards, specifications and codes of practice covers concrete mix design, manufacturing, transportation and placing. The deterioration of concrete is similarly well researched and documented.

Concrete is particularly at risk from chlorides and sulphates in the soil and groundwater, and carbon dioxide within groundwater and atmospheric air. Barry has listed more than 50 chemicals that may be considered potentially aggressive to concrete (Table A5.1).

Table A5.1 *Some of the substances potentially corrosive to concrete*

Substance	Comment
Inorganic acids	especially sulphuric, nitric and hydrochloric
Metal oxides	small amounts of zinc or lead oxide retard hardening of OPC concrete
Ammonium nitrate	
Coal and clinker	particularly the wet oxidation of pyrites to sulphate
Chlorides	especially corrosion of reinforcing steel
Sulphates	primarily chemical attack of cement components
Carbon dioxide	reduction of pH of concrete, reinforcing steel becomes vulnerable to corrosion
Beer	
Milk	
Ink	certain types containing acids
Phthallates	alkyl phthallates
Sodium thiosulphate	derived from photographic solutions
Mineral oils	affect the hardening of "green" concrete
Coffee and cocoa beans	attack from organic acids produced by these

As general rule the factors that influence chemical attack of concrete are:

- the concentration of the substance
- the water table, solubility of the substance and mobility of groundwater
- the compaction, cement type and content, type of aggregate, water/cement ratio
- the form of construction.

Effects of carbon dioxide on concrete

The attack on concrete by atmospheric carbon dioxide is called carbonation. Carbon dioxide is present in air to the extent of about 0.03% by volume, and under this pressure is soluble to the extent of 0.00054 g per litre, giving a solution with a pH of 5.7. Some spring waters may contain elevated carbon dioxide dissolved under pressure from underground sources and thus are more acidic. Concrete is potentially at risk from acidic groundwater from any source. Since these are natural sources they are not considered further here.

Sulphate attack on concrete

Sulphur is an essential element for plant growth. Sulphates are present naturally in many soils, predominantly, in the UK, London Clay, Lower Lias, Oxford Clay and Weald Clay. Sulphur species may also be present in soils due to contamination by industrial processes, which includes concentration due to mineral extraction such as mining or combustion of coal. Other high-sulphate material includes some blastfurnace slags and oil shale residues. Even old bricks may contain elevated sulphate, including those with gypsum plaster attached.

The general mechanism of sulphate attack is to react with the tricalcium aluminate, tetracalcium aluminoferrites and calcium hydroxide within the cement. These reactions are expansive and cause breakdown in the structure of the concrete. They depend on the nature of the sulphate present, with magnesium then sodium sulphates being rather more aggressive than calcium sulphate related to the individual compound's solubility. Guidance on the placement of concrete is sulphate-bearing soils is given in BRE (1996).

The thaumasite form of sulphate attack on concrete

Thaumasite occurs naturally as a rare mineral in altered basic rocks and altered limestones. Over the past 30 years thaumasite has been found as a deterioration product within several cement-based building structures that had been identified as suffering from sulphate attack. The thaumasite form of sulphate attack (TSA) has been observed recently in bridge foundations on the M5 motorway, and has also been documented in columns supporting a building in arctic Canada, tunnel linings, sewage pipes and road sub-bases. TSA is significant because, in a broadly similar fashion to more common forms of sulphate attack, it can cause buried concrete gradually to lose strength; sometimes the affected areas expand in volume. The attack generally begins at the outer margins of the concrete and works progressively inwards. In due course the matrix of the concrete may disintegrate into a soft paste. TSA is not simply confined to below ground concrete; evidence of TSA has been found in the following situations:

- sulphate-bearing brickwork – TSA has been observed in mortars and renders in contact with sulphate-bearing clay bricks
- concrete bricks made with aggregates contaminated with sulphides/sulphates
- historical mortars
- concrete used for floor slabs
- gypsum and lime gypsum plasters in contact with cement renders.

The mechanisms of the Thaumasite attack process and the factors affecting rate of deterioration are the subject of study, however several parameters have been provisionally identified as influencing the rate of deterioration:

- the availability of sulphate ions at the ground/concrete interface
- the quality of the concrete (increased rate with lower cement content, higher water/cement ratio and higher porosity)
- the amount of carbonate especially in the finer fractions within the concrete (increased rate with greater amount of fine carbonate
- the temperature of exposure – low temperature (less than 15°C is generally conducive to Thaumasite formation).

Thaumasite cases documented to date are considered to have occurred largely due to naturally derived "contamination". However research is ongoing to consider the effects of disturbance of ground with elevated sulphides, and the those of mine and industrial wastes rich in sulphate and sulphide. Current research indicates that the disturbance of soil rich in sulphides but low in sulphate can cause increase in the sulphate content by sulphide oxidation.

Effects of chloride on concrete

Chloride is an important agent in the deterioration of concrete. The most common result of exposure of concrete to chlorides is corrosion of the reinforcing steel leading to expansive forces, then cracking and eventual breaking away of pieces of the concrete in contact with the steel (spalling). Note that observed corrosion of the reinforcement is one of a number causes of the observed deterioration of concrete. Other promoters of deterioration include structural overload, impact and abrasion, shrinkage or expansion and other chemical attack or interaction such as sulphate attach or alkali-aggregate reaction.

The chloride may be derived from natural sources, such as seawater and salt-laden air, or by the use of de-icing salts on roads. The chloride may be present within aggregates used for concreting (and therefore "built in" to the concrete) or may be present due to industrial contamination. Chlorides are likely to be encouraged to migrate in conditions of continual wetting and drying. The presence of oxygen and sufficient quantities of free chloride ions in the pore water of concrete can produce concrete corrosion, even in the presence of the highly alkaline conditions of good quality concrete, and pitting of the steel occurs. In favourable conditions, the pits can rapidly expand to significant depths.

Risks due to chloride attack are reduced by careful control of the availability of free chloride ion in the vicinity of reinforcing steel. Levels of chloride within concrete are controlled by reducing the amount of "built in" chloride, or by reducing the amount of externally derived chloride from reaching the reinforcing. Several publications are recommended below for further reading on this subject.

A5.2.3 Effect on masonry

Masonry, including brickwork (bricks, mortar) and stonework, is susceptible to deterioration in aggressive environments. The durability of brick and stonework in aggressive environments is related principally to its porosity. A low porosity is inductive of greater durability (Barry, 1983), as is evidenced, for example, in some Victorian era infrastructure made using good quality engineering grade brick. The contamination induced deterioration of brick and stone will depend on the chemical substances to which they are exposed. The deterioration may result either from chemical interaction or from expansive forces generated by the build-up and crystallisation of salts within pores. Salts are transported by water to the interior of the brick or stone structure.

Masonry mortars often consist of building sands mixed with either cement or lime or both. Mortar tends to be susceptible to attack from similar chemical species to those that attack concrete. In particular, sulphate attack is common, where sulphates have either arisen from within the bricks (the sulphate content of modern bricks is carefully controlled) or from the external environment.

A5.2.4 Effect on rubbers and plastics

Plastics and rubbers are complex hydrocarbon molecules. The term "plastic" refers to a wide range of polymers. Aggressive ground may have an adverse effect on plastics causing them to deteriorate. The deterioration may involve lysis of the bonds linking the individual molecules of the polymer or may involve reactions in any side chains such as those leading to cross-linking (cross-linking is the chemical bonds formed between adjacent molecules). In effect all this means a change to the physical properties of the polymer. The changes may include swelling, cracking, dissolution and microbial degradation. Rubbers are cross-linked polymers that also contain other substances such as fillers and other additives. Rubber deteriorates due to swelling, cracking and hardening and microbial degradation.

As a general rule, plastics are particularly at risk from hydrocarbon and chlorinated hydrocarbon solvents, and extremes of pH. Rubbers are degraded principally by oxidation, for example by oxidising acids and salts such as nitric acid and metal salts, but are resistant to strong acids and alkalis. Rubber is also susceptible to attack by certain hydrocarbon species. Primarily the risk assessor should refer to the manufacturer's technical literature to determine whether there is likely to be an adverse effect on plastics intended for placement or in actual use within aggressive ground. Paul (1994) provides useful background on the performance of plastics and rubbers in aggressive ground.

A5.2.5 HOW TO ASSESS THE RISKS TO CONSTRUCTION MATERIALS

Effects on construction materials and structures are rarely noticeable instantaneously, so the risk assessment process aims to estimate the effects of aggressive ground conditions in the long term.

Where generic guidance exists (such as BRE (1996) for sulphates), this should be used as a primary screening tool. It is commonly found that guidance of this sort is not directly available, however. The assessor may turn to data published on the effects of contaminants on durability, obtained by laboratory testing. This may be conservative, since the data may have been obtained under extreme conditions that were not representative of the site. Nevertheless, it does provide an indication of the potential effects that may be encountered. A conservative approach may also allow for less than ideal standards of workmanship that may influence durability (especially concrete). Most guidance does not necessarily come with determined thresholds "below which effects are unlikely". The assessor should consider whether the concentration of the contaminant of concern is sufficiently low that long-term difficulties are unlikely.

In extreme cases the assessor may commission special tests on the construction material to determine the effects of its exposure under controlled conditions to substances that will be encountered on the site. This is only likely to be possible when there is:

- sufficient time within the assessment programme (often dependent on the site development programme)
- the tests can be made sufficiently representative of the site conditions to be comparable.

The data on the durability of materials is largely qualitative, so professional judgement is needed to appraise risks. The assessment may conclude that a meaningful assessment of effects on the material or structure cannot be made due to the lack of appropriate data. The recommendation would be for over-design to compensate for perceived reduction in the integrity of the material or structural element. Risk control design measures might include specification of a higher grade of concrete, or substitution of a component made of one material for an equivalent made of different material. An alternative is to include a sacrificial layer of the material, which would deteriorate during the design life of the structure without affecting the overall integrity and protective coatings or barriers.

A5.3 FURTHER READING

BAMFORTH, P B, PRICE, W F and EMERSON, M (1997)
An international review of chloride ingress into structural concrete, Contractor Report 359
Transport Research Laboratory, Crowthorne, UK

BARRY, D L (1983)
Material Durability in Aggressive Ground, Report 98
CIRIA, London

BRE (1996)
Sulphate and acid resistance of concrete in the ground, Digest 363
Building Research Establishment, Watford, UK

BRE (2000)
Corrosion of Steel in Concrete, Digest 444, Parts 1 to 3
BRE Centre for Concrete Construction, Watford, UK

CRAMMOND, N J and DUNSTER, A M (1997)
Avoiding Deterioration of Cement Based Building Materials
Lessons from Case Studies 1
BRE, Watford

DETR (1999)
The Thaumasite Form of Sulfate Attack: Risks, Diagnosis, Remedial Works and Guidance on New Construction, Report of the Thaumasite Working Group for DETR, January 1999
HMSO, London

EGLINGTON, M S (1987)
Concrete and its Chemical Behaviour
Thomas Telford, London.

ENVIRONMENT AGENCY (in preparation)
"Guidance on the Risks of Contaminated Land to Buildings, Building Materials and Services", Project P5-035

HEWLETT, P C (1999)
Lea's Chemistry of Cement and Concrete, 4th edition
Wiley, London

PAUL, V (1994)
Performance of Building Materials in Contaminated Land, Report 255
Building Research Establishment, Watford, UK

VASSIE, P R (1979)
Influence of Chlorides on the Alkalinity of Concrete, Laboratory Report 915
Transport Research Laboratory, Crowthorne, UK

CIRIA (2000)
Concrete technology for cast in-situ foundations, Funders Report CP/68
CIRIA, London

A6 Risk assessment software models

Table 6.2 provides a summary of the principles and methodologies of selected risk assessment models packages.

A6.1 BP RISC

Risk Integrated Software for Cleanups (RISC) has been developed for BP over the last eight years, and is subject to continued development. It is a Windows-based package that incorporates a range of tools developed to assess a wide range of exposure scenarios. In brief, the model may be used to:

- assess the potential for adverse human health impacts due to exposure to contaminated soil, water and air
- calculate target clean up levels for soils and groundwater
- estimate contaminant transport between soils, groundwater and air.

The model provides these assessments based on several different tools, comprising:

- a stand-alone spreadsheet that provides initial risk screening using ASTM E1739-95 RBCA Tier 1 algorithms
- a range of tools to evaluate a wide range of exposure scenarios and transport pathways (for example, ingestion/inhalation via groundwater/indoor air from soil or groundwater sources)
- the ability to carry out forward and backward calculations, ie:
  - assess potential adverse health affects arising from user-specified soil, groundwater or air concentrations (the forward calculation), or
  - to estimate clean-up levels based on pre-determined acceptable potentials for adverse health effects
- a means (for different scenarios) to distinguish between the presence of miscible/immiscible phases in the source
- pre-defined exposure parameters (eg for residential child, adult etc)
- probabilistic analysis and site data statistical analysis tools
- simultaneous multiple pathway and multi-receptor analysis.

Input data requirements are fairly typical, including source size and geometry, source media properties (eg soil density, porosity, thickness etc), contaminant-specific properties (concentrations, partition coefficients etc) and exposure reference values. Default values for different soil types are available and chemical properties can be retrieved from the contaminant database.

Depending on the level of analysis undertaken, and the mode of analysis, a range of output options is available including statistical tables, graphs and charts.

As with all models of this type, the user needs to be aware of the various algorithms used in model calculations and take care that the assumptions used in both model calculations (and default parameters where used) are appropriate with the site-specific conditions for the model application. At least basic knowledge of the fundamentals of contaminant migration, exposure and risk assessment is necessary to ensure correct choice of input parameters.

BP RISC can be obtained from:

Environmental Software Online, LLC
520 Chicopee Row, Groton, MA 01450, USA
Tel: (978) 448-5818; fax: (978) 448-6280
Website: www.groundwatersoftware.com; email: info@groundwatersoftware.com

A6.2 CONSIM

CONSIM is a Windows-based probabilistic contaminated land risk-assessment model that allows the user to assess the risk posed to groundwater by leaching contaminants. Golder Associates developed the model on behalf of the Environment Agency.

Using available site investigation information, and a range of default or user-entered parameters, CONSIM may be used to model contaminant mobilisation and transport through the water environment. The probabilistic elements of the model are generated through a Monte Carlo simulation of input parameter ranges, and model outputs are in graphical or statistical format. Retardation and first order biodegradation can be included if required.

The model may be used to assess the:

- potential for pollution of controlled waters
- need for additional site investigation data to quantify the risk to groundwater
- concentration of contaminants (in the water environment) at the source, at the base of the unsaturated zone and within the aquifer and at a receptor.

The model is based on a staged (tiered) structure, in which the extent of analysis and data demand increases with each stage. Within CONSIM, the stages represent progression of the analysis through defined parts of the water environment, such that:

- Level 1 includes for the assessment of contamination of the water environment (ie pore water) in the immediate vicinity of the source
- Level 2 considers the above, plus contamination of the water environment at the base of the unsaturated zone
- Level 3 considers all the above, plus an assessment of the travel time to, and concentrations at, a user-defined receptor.

At each level the contamination concentrations of each selected contaminant are compared to water quality standards. These standards may be user-selected default values (such as the Water (Water Quality) Supply Regulations 1989), or may be user-defined site-specific values. At the end of each level of assessment, where these standard values are not exceeded, it can be assumed, broadly speaking, there is no risk to the water environment, and no further level of assessment is warranted. This does depend, to some extent, on the objectives for carrying out the assessment.

It should be noted that the "levels" in the CONSIM model do not relate directly to the tiered structure of either the Model Procedures or the EA Methodology for the derivation of Remedial Targets for Soils and Groundwater to Protect Groundwater Resources. However, the model itself provides one mechanism for carrying out Tiers 3 and 4 of these procedures.

Typically, model input requirements are as detailed in Table A6.1.

Table A6.1 *Model input requirements*

Model level	Model data categories	Typical data set needed
1, 2, 3	Contaminant inventory	Soils quality or leachate test data
1, 2, 3	Contaminant detail	Unsaturated zone partition co-efficients
2, 3	Infiltration	Site infiltration rate (mm/yr)
2, 3	Source	Source dimensions, soil density, porosity
2, 3	Unsaturated pathway	Depth to saturated zone, porosity, density, hydraulic properties, dispersivity
2, 3	Aquifer pathway	Flow path dimensions, mixing zone thickness, density, hydraulic conductivity and gradient, porosity, dispersivity
3	Receptor	Receptor locations etc

The initial levels of the model are not data intensive, however, data requirements increase through each level. Defaults set in the model (eg for dispersivity, partition co-efficient etc) are useful and contaminant-based. Users should be aware of the assumptions employed not only in the default data sets but also in the algorithms (and concepts behind these) for deriving various output data, particularly in Levels 2 and 3.

Presently, the program cannot incorporate actual (site-measured) water quality data, but derives water environment concentrations via determination of contaminant partitioning during infiltration through both soil and unsaturated zone. Where significant actual water quality data are available, use of groundwater contaminant transport models should be considered to evaluate contaminant concentrations at potential receptors.]

ConSim can be ordered from the ConSim Helpdesk at:

Golder Associates (UK) Ltd
Landmere Lane, Edwalton
Nottingham NG12 4DG
Tel: 0115 945 6544; fax: 0115 945 6540
Email: consim@golder.com

A demonstration version can be downloaded from the Environment Agency website www.environment-agency.gov.uk.

A6.3 RBCA TOOLKIT FOR CHEMICAL RELEASES

The RBCA Tool Kit for Chemical Releases was designed by Groundwater Services Inc. It is based upon ASTM PS-104, *Standard Provisional Guide for Risk-Based Corrective Action* (ASTM, 1998). The risk assessment procedure is reported to be consistent with current USEPA guidelines.

The model completes all the calculations required for a Tier 1 and 2 RBCA analysis. It also calculates baseline risk level and risk-based clean-up standards. Several fate and transport models are also included.

The exposure pathways considered are:

- groundwater exposure
- surface water receptors
- surface soil exposure
- air exposure.

The groundwater exposure pathway assesses direct ingestion of groundwater or exposure to surface water affected by contaminated groundwater. The groundwater can be either directly contaminated or through leaching from a contaminated soil. Three basic types of receptor are considered, residential, commercial and MCL (maximum contaminant level). The MCL allows for clean-up concentrations to be calculated based on water quality objectives at the receptor location.

The surface water pathway assesses exposure to humans (through swimming), humans (through fish consumption) and direct exposure of aquatic life. The surface water pathway involves contamination of surface water by contaminated groundwater. The surface water exposure scenario involves direct contact with soils (dermal and ingestion). Two receptors are specified: residential and commercial. An option to assess the "construction worker" receptor is included.

The air exposure pathway includes both indoor and outdoor air inhalation of vapours that emanate from contaminated soils/groundwater. Three receptors are specified, residential, commercial and time-weighted average. The latter allows for the exposure limits to be set to permissible occupational exposure limits. An option to assess the "construction worker" receptor is included (for outdoor air inhalation).

The default exposure pathways within the model correspond to reasonable maximum exposure (RME) values specified by USEPA (USEPA, 1991). These are fully adaptable to actual site exposure scenarios. The model contains a chemical database with more than 100 compounds, including TPH fractions. Physical and toxicity data have been compiled from many sources, but mainly ASTM and USEPA. All are fully referenced within the model. The database is fully customisable. The ability to calculate basic statistics (mean, maximum, upper confidence levels) is included within the model.

There are several contaminant transport models included:

- ASTM surface soil volatilisation model
- Johnson-Ettginger volatilisation model for soils and groundwater
- ASTM leachate soil model
- Soil Attenuation Model (soil to groundwater leaching)
- Domenico Model (groundwater dilution).

It is possible to replace the values generated by these models with user specified parameters from either measured or other model outputs. The toolkit also allows for both steady state contaminant source and a decaying contaminant source.

RBCA Toolkit is available from:

Groundwater Services Inc
2211 Norfolk, Suite 1000
Houston, Texas 77098-4044
Tel (713) 522-6300; fax (713) 522-8010
Website: www.esnt.com; email admin@www.esnt.com

A6.4 RISC-HUMAN

RISC-HUMAN (V3.0) has been developed by the Van Hall Institute. It is based upon the CSOIL model (as used to calculate the Dutch Intervention Levels).

The software models risks to humans from contaminated soils, surface water, groundwater, vegetables, fish, meat and milk (several of these pathways are new for this version of the model). The exposure routes and default values of parameters are valid for a residential area with gardens. The model assumes a constant contaminant source.

RISC-HUMAN can assess concentration of contaminants within the following media:

- outdoor air
- indoor air
- groundwater
- soil
- surface water
- milk, fish, vegetables, and meat.

The model also contains several "exposure scenarios". These are:

- drinking water
- showering
- indoor air
- outdoor air
- swimming
- dust and soil indoors
- dust and soil outdoors
- consumption of vegetables, fruit, dairy products, meat and fish.

Unlike Risk* Assistant, RISC-HUMAN includes dermal exposure to soils.

RISC-HUMAN can be obtained from

Van Hall Instituut Business Center
Agora 1, PO Box 1754
8901 CB Leeuwarden
The Netherlands
Tel: +31 58 2846160; fax: +31 58 2846199
Website: www.vhall.nl, email: Risc.Human@pers.vhall.nl

A6.5 RISK* ASSISTANT

Risk* Assistant was developed by the Hampshire Research Institute of Alexandria, Virginia, USA, with assistance from US Environmental Protection Agency. Fundamentally it is based upon the USEPA risk assessment methodology (USEPA, 1989) and uses toxicity information from published toxicological data sources, including USEPA's Integrated Risk Information System (IRIS) and Health Effects Assessment Summary Tables (HEAST) databases, New Jersey Hazard Data and California Hazard Data. The facility exists to use a customised database instead.

Risk* Assistant allows the risks to humans posed by contaminated media to be estimated.

Risk* Assistant addresses contamination within the following environmental media:

- air
- groundwater
- surface water
- soil
- sediment
- fruit and vegetables
- fish, meat and dairy products.

To an extent the model predicts transfers between these media. The model also allows for statistical analysis of contaminant concentration data, for example, the calculation of upper confidence limits.

The model uses reference doses (RfD) to assess non-carcinogenic risks and slope factors to assess carcinogenic risks.

The model contains default information on average and sensitive receptors ("populations"), which are based upon American statistics. It is possible to customise these data sets, however.

The model also contains several exposure scenarios. These are:

- drinking water
- showering
- indoor air
- outdoor air
- swimming
- dust and soil indoors
- dust and soil outdoors
- consumption of vegetables, fruit, dairy products, meat and fish.

It is worth noting that the dust and soil (indoors and outdoors) exposure scenario does not account for dermal contact with contaminated dust/soil.

The software package also contains copies of the Win ISC2 (LT) model and Stream*Model to model contaminant migration in air and water respectively.

Risk*Assistant can be obtained from:

Thistle Publishing Inc
PO Box 1327, Alexandria
VA 22309, USA
Fax: 703/684-7704
Website: www.thistlepublishing.com; email: jhoway@hampshire.org

A7   Case studies

A7.1   CASE STUDY 1 – HOUSING DEVELOPMENT, LIVERPOOL, UK

Objectives of the risk assessment

To assess the risks posed by redevelopment of former industrial premises for residential housing.

Final scope of risk management process

Desk study, invasive site investigation, remediation of contamination derived from industrial sources including some radioactive contamination.

Why was contamination suspected

The site had a long history of industrial usage by number of potentially contaminative industries.

Site setting and Phase 1 studies

The site area was 6.1 ha, of which 5.3 ha were derelict and 1.3 ha occupied by light industry. The site geology consisted of made ground overlying sand, in turn overlying boulder clay with occasional peat deposits.

A historical study was conducted to aid decision-making. It revealed that past uses included chemical works, manufacture of tanning extract and paints, timber yard, timber treatment area and railway sidings. Particular constraints were identified.

Particular environmental constraints

The site was located adjacent to existing residential development. Redevelopment would have to be undertaken to minimise migration of contaminants towards this development.

Action resulting from Phase 1 study

Invasive site investigations were carried out to determine actual contamination present, its nature and extent, and the potential impact on future residential development.

Four site investigations were conducted, involving excavation of boreholes, trial pits, sampling of soils and groundwater, and monitoring of soil gases within standpipes. Some radioactivity testing was carried out on an area where tin slag had been deposited.

The investigations demonstrated that most contamination is localised into one of three areas. Widespread metals contamination was found, little organic contamination (for example mineral oils), radioactivity associated with tin slag (gamma radiation), elevated concentrations of soil gases (carbon dioxide and methane).

Generic assessment criteria used

ICRCL values were used as screening values. For radioactivity, criteria derived from International Commission for Radiological Protection (ICRP).

Problems encountered and consequences

The radioactivity within the tin slag had not been anticipated. Some of this material had been used earlier for sub-base material for roads within the adjacent residential development. Subsequently, further risk assessment was directed towards these off-site areas. The discovery of radiation caused significant delays while remediation action was taken; consequently, costs increased significantly.

Lessons learnt

The risk assessment, especially at site investigation stage, should address the possibility that radiation (or any contamination requiring special handling procedures) might be present, and allow programme time and budget accordingly.

A7.2 CASE STUDY 2 – CONTAMINATION OF GROUNDWATER AT MAINTENANCE FACILITY

Objectives of the risk assessment

A study was commissioned of a disused vehicle maintenance facility in connection with possible redevelopment for residential housing.

This case study is intended to illustrate the importance of identifying all the potential pathways and the importance of a thorough Phase 1 desk study.

Final scope of risk management process

Desk study, invasive site investigation, remediation of contamination derived from industrial sources.

Why was contamination suspected

The site had comprised a collection of warehouse buildings used over many years for various purposes, but predominantly for vehicle maintenance and refuelling.

Site setting and Phase 1 studies

The layout consisted of several maintenance buildings, positioned around a main workshop. The (Phase 1) study of historical maps revealed that four oil storage tanks had formerly been located near to the main workshop. Studies of geological plans indicated that the site geology comprised (in sequence):

- made ground (mainly ash sand and clay)
- Terrace Gravels
- Woolwich and Reading Clays
- Thanet Sand
- Chalk.

The site hydrogeology was also examined. The Terrace Gravels were identified as water-bearing (upper aquifer) with further groundwater within the Thanet Sand and underlying chalk (lower aquifer). The lower aquifer is confined by the overlying Woolwich and Reading Beds.

A conceptual model was developed as a basis for the risk assessment. The potential contaminant-pathway-receptor scenarios examined included the possibility that contamination from spills or leaks from the former oil tanks had migrated vertically to enter both the upper and lower aquifers. The potential pathway to the lower aquifer would be via fissures within the clay or existing via piled foundations. Either of these could provide a conduit for contaminants to impact the lower aquifer.

Particular environmental constraints

None identified.

Action resulting from the Phase 1 study

A Phase 2 site investigation was planned and instigated. Unexpectedly, a disused water well was discovered within the works close to the oil tanks. This well had not been identified during the desk study. The well had been loosely capped and was found to penetrate the Chalk. The well was filled with oily water, and in addition it was found that some surface drains were found to discharge into the well, presumably after the well was abandoned.

Samples for groundwater analysis were taken from boreholes installed during the site investigation. Laboratory analysis indicated that groundwater from the Upper aquifer was contaminated with diesel oil and polyaromatic hydrocarbons (PAH) in the vicinity of the tanks. A thin layer of oil (free product) on the groundwater of the lower aquifer was noted in the vicinity of the former tanks.

Laboratory analysis indicated that water quality in the lower aquifer was improved over the upper although concentrations of diesel oil and PAH were elevated in some boreholes, principally down-gradient from the oil tanks. Groundwater concentrations in the upper and Lower aquifers were plotted on a site plan. The highest concentrations in the upper aquifer were found in the vicinity of the oil tanks. In the lower the highest concentrations were in the vicinity of the well. It was concluded that the disused well was a suitable pathway for contaminant migration, in addition to any old foundation piles that may already have penetrated the Woolwich and Reading Beds.

Generic assessment criteria used

Primarily "Dutch" criteria for soils and groundwater, supplemented by ICRCL and other criteria as required.

Lessons learnt

In conclusion, leaks and spills from the former oil tanks apparently contributed contaminants (diesel oil and PAH) to the upper aquifer, and to a lesser degree the lower aquifer, due to downward migration via the well and to a lesser extent fissures within the clays. This pollution would continue until the sources (in this case the free product) were exhausted. The contamination sources across the majority of the site effectively ceased once the railway engineering facilities were closed. Pollutants would migrate down the hydraulic gradient and may potentially move off-site at some point in the future, although monitoring of boreholes outside of the site boundary indicated that there is no evidence at the moment for off-site migration.

In this example, careful planning and execution of the site investigation maximised the discovery of the buried well. In fact, had the well not been found, careful interpretation of the groundwater monitoring data and laying out of the results on a groundwater "contour plan" would have indicated that a large hot spot of contamination existed close to the tanks in both the upper and lower aquifers. A significant contaminant pathway between the two aquifers could be expected, therefore. Had the well not been found during the main investigation, a second phase of investigation would have been recommended to prove the existence of the pathway. This would have concentrated on the centre of the hot spot, as a result of which the well is likely to have been discovered.

A7.3 CASE STUDY 3 – LAND PURCHASE BY HOLDING COMPANY, NORTH-WEST ENGLAND

Objectives of the risk assessment

To determine whether contamination was present at a site to enable a potential purchaser to assess the likely extent of his future liabilities.

Final scope of risk management process

Historical studies, site walkover, employee interviews and preliminary risk assessment.

Why was contamination suspected

The site had been used for over one hundred years for chemicals manufacture. The current owner had recently ceased chemical manufacture at the site.

Site setting and Phase 1 studies

Studies were conducted that included reviews of historical documentation, a site walkover visit and interviews with former employees at the site. No formal site investigation has been undertaken at the site, however considerable contamination was visually identified in inspection pits. In addition there were accounts by ex employees of former site waste disposal practices that indicated use of soakaways for liquid waste, reuse of demolition materials to fill in voids and general dust fallout that will have contributed contamination to the near surface. Contamination was therefore thought to be widespread and to include as a minimum metals, inorganic species, hydrocarbons. asbestos, bulk raw materials, product and wastes and minor stored materials (individual containers/drums). This contamination was anticipated within ground and within some buildings (such as residues of hazardous materials within drains and traps).

The geology of the site consists of made ground underlain by alluvium, in turn underlain by Clay and Sandstone

Particular environmental constraints

The former Works Water Well, placed into the sandstone underlying the site and disused after around 1959 has been identified as a possible conduit for the vertical migration of contamination within the near surface ground to lower levels, potentially impacting upon the underlying aquifer. The current status of the well with regard to sealing and capping after being taken out of use was unknown, as is its exact location. The site was located adjacent to a stream.

Action resulting from Phase 1 study

No further technical action was required. The potential purchaser used this information to negotiate a sale deal with the site owner that included a sizeable reduction in the purchase price in return for the purchaser taking on responsibility for the contamination. The reduction resulted from agreement by both parties that contamination was likely to be extensive (although this had not been proved by investigation) and extensive remedial works would be necessary to render the site acceptable for commercial development.

A7.4 CASE STUDY 4 – LARGE BROWNFIELD SITE REDEVELOPMENT, ESSEX, UK

Objectives of the risk assessment

To identify a remediation process that would provide all stakeholders with confidence in the redevelopment of a former explosives factory site for residential housing, commercial development and parkland.

This case study is intended to illustrate the impact of stakeholder confidence on remediation design derived from Phase 1 and 2 risk assessments.

Final scope of risk management process

Desk study, invasive site investigation, implementation of a £6.5 million remediation process, confirmatory investigation to demonstrate success of remediation.

Why was contamination suspected

The site had been used for about 100 years for the research, development, manufacture and testing of explosive materials. Some 200 specialist buildings were present on the site, as well as a network of canals backfilled during the site's operational history.

Site setting and Phase 1 studies

The site area was 100 ha. Completion of a Phase 1 study indicated the potential for extensive ground contamination, but suggested possible cost-effective remediation based on moving all possibly contaminated materials beyond the potentially redevelopable areas. This scenario was welcomed by the local authority and potential developers of the site, as it overcame the local perception of unacceptable risks associated with development over land formerly used for the manufacturing of explosives.

Particular environmental constraints

The site's history as an explosive factory made perceived risk a major factor in remediation design and execution. Local pressure groups were convinced that the site was significantly contaminated and could not be appropriately remediated.

The acute point-source risk from explosive articles could not be assessed by intrusive investigation, since although detection techniques can be used for munitions, the items to be detected in this case included types generally too small to be detected by these techniques. Thue ground could not be sampled to provide robust data in this respect.

Action resulting from Phase 1 study

An outline planning permission for 19 ha of commercial development and 11 ha of housing was obtained (based on the Phase 1 study), with the rest of the site (which was within the Green Belt) to become parkland. This permission was conditional upon the satisfactory execution of works covered by a separate conditional remediation planning permission. This permission required a comprehensive Phase 2 investigation to be carried out, plus the preparation and execution of a detailed remediation plan. Each stage was to be agreed in advance with the local authority, which reserved the right to seek advice from the Environment Agency.

The Phase 2 investigation demonstrated that contamination levels across most of the site were light, and on a conventional site would have required relatively small volumes to be treated/removed. However, for the purposes of stakeholder confidence, and to overcome the issue of point-source acute explosive risk, a much more intensive remediation plan was designed.

The remediation plan agreed and executed was for the full depth of fill present on the development site to be relocated to the parkland area, and capped over with clean site-won materials. The redevelopment area was then subjected to a confirmatory investigation, and subsequently regraded. Materials agreed with the local authority to be significantly contaminated were placed in a waste management-licensed landfill constructed on part of the parkland area.

Generic assessment criteria used

For conventional contaminants, ICRCL assessment criteria were used, some modified with the agreement of the local authority to take account of elevated background levels in the area. For explosives, acceptable levels were determined and agreed with the local authority based on guidance produced internally by Royal Ordnance.

A major contaminant of concern on the site was asbestos (from steam pipe lagging) in the soil. The local authority set risk assessment and clean-up targets based on the Harry Stanger analysis method, which categorises levels from Cat 1 (>0.1%) to Cat 4 (<0.001%).

Problems encountered and consequences

The planning period before the start of the works became protracted because of the complex nature of the contamination and the requirements for investigation, assessment and remediation, also because of interaction with other local development and infrastructure. There was considerable caution within the local authority over appropriate risk assessment and remediation, given the special problems of the site. However, the long planning period allowed ample time for technical issues to be resolved and no significant environmental problems occurred during the project.

Lessons learnt

Complete involvement of the local authority at all stages of the project proved to be essential. Stakeholder risk perception proved to be a more important factor for remediation design than the results of the Phase 2 investigation.

A7.5 CASE STUDY 5 – SMALL HOUSING DEVELOPMENT, UK

Objectives of the risk assessment

To identify a remediation process that would enable a housing developer to continue with house construction on a site where a landfill gas problem was identified late in the construction planning cycle.

This case study is intended to illustrate some of the potential pitfalls in the redevelopment planning process associated with risk assessment.

Final scope of risk management process

Desk study, invasive site investigation, implementation of a £350 000 remediation process, confirmatory monitoring to demonstrate success of remediation.

Why was contamination suspected

The site had formerly been used as a vehicle maintenance depot. A Phase 1 investigation and limited Phase 2 work had been performed for the vendor of the site. The work indicated no significant contaminative issues, although they did identify that part of the site had been a gravel pit which had subsequently been infilled.

The purchaser of the site (a housing developer) conducted a limited supplementary Phase 2 (primarily geotechnical) site investigation, and discovered significant landfill gas levels being generated from the in-filled gravel pit.

Site setting and Phase 1 studies

The site occupied 3 ha. The Phase 1 study conducted for the vendor failed to identify the potential for landfill gas problems. Further intrusive investigation by the buyer, primarily carried out for foundation design purposes, had identified potentially significant landfill gas-related problems on the site. The buyer therefor commissioned more thorough Phase 1 and Phase 2 investigations. These identified minor and localised ground contamination originating from the former vehicle servicing activity, along with significant levels of landfill gas generated from a layer of organic material at the base of the in-filled gravel pit area, which extended beyond the site boundary.

Particular environmental constraints

The source of the landfill gas extended beyond the boundaries of the site.

Action resulting from Phase 1 study

A detailed planning permission for the construction of 20 houses had been obtained on the basis of the vendor's Phase 1 risk assessment. This permission was obtained before identification of a potential soil gas issue and made no requirement for site investigation or remediation works. Consequently, the local authority was under no statutory duty to approve the design or implementation of the remediation works. Rather than delay construction of the houses for a potentially extended period while the planning permission was varied, the developer decided to proceed on the basis of non-statutory negotiations with the local authority. Following more comprehensive Phase 1 and 2 investigations, a remediation plan was informally agreed with the local authority and implemented. This required the construction of a cutoff trench to prevent on-site migration of landfill gas, plus excavation of the infilled gravel pit. The organic sludge was removed off site, but the remaining excavated material was used as backfill.

Generic assessment criteria used

ICRCL assessment criteria for ground contamination (with some modifications to reflect local background levels). Waste Management Paper 27 for gas thresholds and protection measures.

Problems encountered and consequences

Because the remediation design seemed robust, and no formal local authority sign-off was required, the developer proceeded with construction and sale of houses prior to a conclusive period of post-remediation gas monitoring. Initial post-remediation monitoring indicated residual gas thresholds at possible problem levels.

It was later shown that no significant evolution of gas was occurring (the levels were apparently due to trace residual organic material within the reused fill). However, the local authority responded in a way that alarmed persons enquiring about buying properties on the site. Some buyers and their legal advisers contacted the local authority for information on the site, and individual local authority officers voiced their concerns over the site to potential buyers. The sales of some properties were then protracted to the discomfort of all parties until sufficient information was available to convince the local authority that the site had been adequately remediated.

Lessons learnt

The absence of any planning conditions in respect of risk assessment or remediation seemed to offer the developer an easy way forward. The building and sale of properties were able to begin sooner than would have been the case if a formal discharge of planning conditions had been required. However, due to disagreement with the local authority over what constituted sufficient evidence that no problem remained, delays to house sales occurred at a crucial point, causing embarrassment to all involved.